Student Solutions Manual

FUNCTIONS MODELING CHANGE

A PREPARATION FOR CALCULUS

Second Edition

Eric Connally

Harvard University Extension

Deborah Hughes-Hallett

University of Arizona

Andrew M. Gleason

Harvard University

et al.

WILEY

John Wiley & Sons, Inc.

COVER PHOTO: ©Nick Wood

 This material is based upon work supported by the National Science
Foundation under Grant No. DUE-9352905. Opinions expressed are
those of the authors and not necessarily those of the Foundation.

To order books or for customer service call 1-800-CALL-WILEY (225-5945).

ISBN 0-471-33382-4

Printed in the United States of America

10 9 8 7 6 5 4

Printed and bound by Courier-Kendallville, Inc.

CONTENTS

CHAPTER ONE

Solutions for Section 1.1

Exercises

1. $w = f(c)$

5. (a) Since $f(x)$ is 4 when $x = 0$, we have $f(0) = 4$.
   (b) Since $x = 3$ when $f(x) = 0$, we have $f(3) = 0$.
   (c) $f(1) = 2$
   (d) There are two x values leading to $f(x) = 1$, namely $x = 2$ and $x = 4$. So $f(2) = 1$ and $f(4) = 1$.

9. Appropriate axes are shown in Figure 1.1.

Figure 1.1

Problems

13. (a) The number of people who own cell phones in the year 2000 is 100,300,000.
   (b) There are 20,000,000 people who own cell phones a years after 1990.
   (c) There will be b million people who own cell phones in the year 2010.
   (d) The number n is the number of people who own cell phones t years after 1990.

17. (a) Figure 1.2 shows the plot of R versus t. R is a function of t because no vertical line intersects the graph in more than one place.
   (b) Figure 1.3 shows the plot of F versus t. F is a function of t because no vertical line intersects the graph in more than one place.

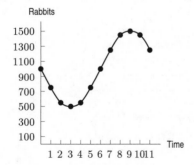

Figure 1.2: The graph of R versus t

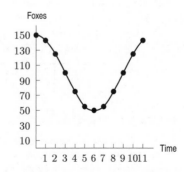

Figure 1.3: The graph of F versus t

(c) Figure 1.4 shows the plot of F versus R. We have also drawn the vertical line corresponding to $R = 567$. This tells us that F is not a function of R because there is a vertical line that intersects the graph twice. In fact the lines $R = 567$, $R = 750$, $R = 1000$, $R = 1250$, and $R = 1433$ all intersect the graph twice. However, the existence of any one of them is enough to guarantee that F is not a function of R.

(d) Figure 1.5 shows the plot of R versus F. We have drawn the vertical line corresponding to $F = 57$. This tells us that R is not a function of F because there is a vertical line that intersects the graph twice. In fact the lines $F = 57$, $F = 75$, $F = 100$, $F = 125$, and $F = 143$ all intersect the graph twice. However, the existence of any one of them is enough to guarantee that R is not a function of F.

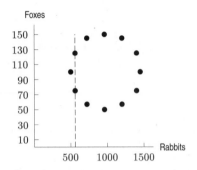

Figure 1.4: The graph of F versus R

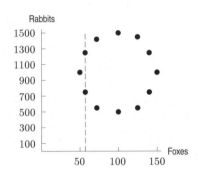

Figure 1.5: The graph of R versus F

21. A possible graph is shown in Figure 1.6.

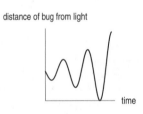

Figure 1.6

25. (a) It takes Charles Osgood 60 seconds to read 15 lines, so that means it takes him 4 seconds to read 1 line, 8 seconds for 2 lines, and so on. Table 1.1 shows this. From the table we see that it takes 36 seconds to read 9 lines.

Table 1.1   *The time it takes Charles Osgood to read*

Lines	0	1	2	3	4	5	6	7	8	9	10
Time	0	4	8	12	16	20	24	28	32	36	40

(b) Figure 1.7 shows the plot of the time in seconds versus the number of lines.

(c) In Figure 1.8 we have dashed in a line to see the trend. By drawing the vertical line at 9 lines, we see that this corresponds to approximately 36 seconds. By drawing a horizontal line at 30 seconds, we see that this corresponds to approximately 7.5 lines.

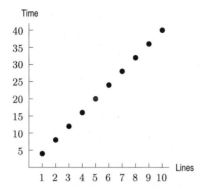

Figure 1.7: The graph of time versus lines

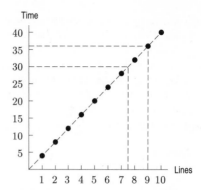

Figure 1.8: The graph of time versus lines

(d) If we let T be the time in seconds that it takes to read n lines, then $T = 4n$.

Solutions for Section 1.2

Exercises

1. **(a)** Let $s = C(t)$ be the sales (in millions) of CDs in year t. Then

$$\begin{array}{c} \text{Average rate of change of } s \\ \text{from } t = 1982 \text{ to } t = 1984 \end{array} = \frac{\Delta s}{\Delta t} = \frac{C(1984) - C(1982)}{1984 - 1982}$$

$$= \frac{5.8 - 0}{2}$$

$$= 2.9 \text{ million discs/year.}$$

Let $q = L(t)$ be the sales (in millions) of LPs in year t. Then

$$\begin{array}{c} \text{Average rate of change of } q \\ \text{from } t = 1982 \text{ to } t = 1984 \end{array} = \frac{\Delta q}{\Delta t} = \frac{L(1984) - L(1982)}{1984 - 1982}$$

$$= \frac{205 - 244}{2}$$

$$= -19.5 \text{ million records/year.}$$

(b) By the same argument

$$\begin{array}{c} \text{Average rate of change of } s \\ \text{from } t = 1986 \text{ to } t = 1988 \end{array} = \frac{\Delta s}{\Delta t} = \frac{C(1988) - C(1986)}{1988 - 1986}$$

$$= \frac{150 - 53}{2}$$

$$= 48.5 \text{ million discs/year.}$$

$$\begin{array}{c} \text{Average rate of change of } q \\ \text{from } t = 1986 \text{ to } t = 1988 \end{array} = \frac{\Delta q}{\Delta t} = \frac{L(1988) - L(1986)}{1988 - 1986}$$

$$= \frac{72 - 125}{2}$$

$$= -26.5 \text{ million records/year.}$$

(c) The fact that $\Delta s/\Delta t = 2.9$ tells us that CD sales increased at an average rate of 2.9 million discs/year between 1982 and 1984. The fact that $\Delta s/\Delta t = 48.5$ tells us that CD sales increased at an average rate of 48.5 million discs/year between 1986 and 1988.

   The fact that $\Delta q/\Delta t = -19.5$ means that LP sales decreased at an average rate of 19.5 million records/year between 1982 and 1984. The fact that the average rate of change is negative tells us that annual sales are decreasing.

   The fact that $\Delta q/\Delta t = -26.5$ means that LP sales decreased at an average rate of 26.5 million records/year between 1986 and 1988.

5. (a) (i) We find the average rate of change in the population as follows. For P_1 from 1980 to 1990,

$$\text{Rate of change} = \frac{\Delta P_1}{\Delta t} = \frac{P_1(1990) - P_1(1980)}{1990 - 1980}$$
$$= \frac{62 - 42}{10} = 2 \text{ thousand people per year.}$$

Thus, P_1 is growing, on average, by two thousand people per year. For P_2 over the same period,

$$\text{Rate of change} = \frac{\Delta P_2}{\Delta t} = \frac{P_2(1990) - P_2(1980)}{1990 - 1980}$$
$$= \frac{72 - 82}{10} = -1 \text{ thousand people per year.}$$

The negative sign tells us that P_2 is decreasing, on average, by one thousand people per year.

(ii) For 1980-1997, the average rate of change of P_1 is:

$$\text{Rate of change} = \frac{\Delta P_1}{\Delta t} = \frac{P_1(1997) - P_1(1980)}{1997 - 1980} = \frac{76 - 42}{1997 - 1980}$$
$$= \frac{34}{17} = 2 \text{ thousand people per year.}$$

That is, the city is gaining 2 thousand people per year. The average rate of change of P_2 is:

$$\text{Rate of change} = \frac{\Delta P_2}{\Delta t} = \frac{P_2(1997) - P_2(1980)}{1997 - 1980} = \frac{65 - 82}{1997 - 1980}$$
$$= \frac{-17}{17} = -1 \text{ thousand people per year.}$$

That is, the city is losing a thousand people per year.

(iii) For 1985 to 1997, we have:

$$\frac{\Delta P_1}{\Delta t} = \frac{76 - 52}{1997 - 1985} = \frac{24}{12} = 2 \text{ thousand people per year.}$$

That is, the city is gaining 2 thousand people per year. The average rate of growth for the second population is:

$$\frac{\Delta P_2}{\Delta t} = \frac{65 - 77}{1997 - 1985} = \frac{-12}{12} = -1 \text{ thousand people per year.}$$

That is, the city is losing a thousand people per year.

(b) The average rate of change of each population is the same on all three time intervals. Each population appears to be changing at a constant rate. The first population is growing, on average, by 2 thousand people per year in each time interval. The second population is dropping, on average, by 1 thousand people per year in each time interval.

Problems

9. (a) (i) We have

$$\frac{f(2) - f(0)}{2 - 0} = \frac{16 - 2^2 - (16 - 0)}{2} = -\frac{4}{2} = -2.$$

This means $f(x)$ decreases by an average of 2 units per unit change in x on the interval $0 \leq x \leq 2$.

(ii) We have

$$\frac{f(4) - f(2)}{4 - 2} = \frac{16 - (4)^2 - (16 - 2^2)}{2} = \frac{-16 + 4}{2} = -6.$$

This means $f(x)$ decreases by an average of 6 units per unit change in x on the interval $2 \leq x \leq 4$.

(iii) We have

$$\frac{f(4) - f(0)}{4 - 0} = \frac{16 - (4)^2 - (16 - 0)}{4} = -\frac{16}{4} = -4.$$

This means $f(x)$ decreases by an average of 4 units per unit change in x on the interval $0 \leq x \leq 4$.

(b) The graph of $f(x)$ is the solid curve in Figure 1.9. The secants corresponding to each rate of change are shown as dashed lines. The average rate of decrease is greatest on the interval $2 \leq x \leq 4$.

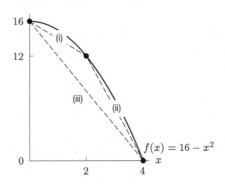

Figure 1.9

13. (a)   (i) Between $(-1, f(-1))$ and $(3, f(3))$

$$\text{Average rate of change } = \frac{f(3) - f(-1)}{3 - (-1)} = \frac{(3^2 + 1) - ((-1)^2 + 1)}{4} = \frac{10 - 2}{4} = \frac{8}{4} = 2.$$

(ii) Between $(a, f(a))$ and $(b, f(b))$

$$\text{Average rate of change } = \frac{f(b) - f(a)}{b - a} = \frac{(b^2 + 1) - (a^2 + 1)}{b - a}$$

$$= \frac{b^2 + 1 - a^2 - 1}{b - a} = \frac{b^2 - a^2}{b - a} = \frac{(b + a)(b - a)}{b - a} = b + a.$$

(iii) Between $(x, f(x))$ and $(x + h, f(x + h))$

$$\text{Average rate of change } = \frac{f(x + h) - f(x)}{(x + h) - x} = \frac{((x + h)^2 + 1) - (x^2 + 1)}{(x + h) - x}$$

$$= \frac{x^2 + 2xh + h^2 + 1 - x^2 - 1}{x + h - x} = \frac{2xh + h^2}{h} = \frac{h(2x + h)}{h} = 2x + h.$$

(b) The average rate of change is different each time. However, it seems to be the sum of the two x-coordinates.

Solutions for Section 1.3

Exercises

1. This table could not represent a linear function because the rate of change of $g(t)$ is not constant. We consider the first three points. Between $t = 1$ and $t = 2$, the value of $g(t)$ changes by $4 - 5 = -1$. Between $t = 2$ and $t = 3$, the value of $g(t)$ changes by $5 - 4 = 1$. Thus, the rate of change is not constant ($-1 \neq 1$), so the function is not linear.

5. The function j could be linear if the pattern continues for values of x that are not shown, because we see that a one unit increase in x corresponds to a constant decrease of two units in $j(x)$.

9. The vertical intercept is 54.25, which tells us that in 1970 ($t = 0$) the population was $54,250$ (54.25 thousand) people. The slope is $-\dfrac{2}{7}$. Since

$$\text{Slope} \; = \; \frac{\Delta \text{population}}{\Delta \text{years}} \; = \; -\frac{2}{7},$$

we know that every seven years the population decreases by 2000 people. That is, the population decreases by 2/7 thousand per year.

Problems

13. **(a)** If the relationship is linear we must show that the rate of change between any two points is the same. That is, for any two points (x_0, C_0) and (x_1, C_1), the quotient

$$\frac{C_1 - C_0}{x_1 - x_0}$$

is constant. From Table 1.25 we have taken the data $(0, 50)$, $(10, 52.50)$; $(5, 51.25)$, $(100, 75.00)$; and $(50, 62.50)$, $(200, 100.00)$.

$$\frac{52.50 - 50.00}{10 - 0} = \frac{2.50}{10} = 0.25$$

$$\frac{75.00 - 51.25}{100 - 5} = \frac{23.75}{95} = 0.25$$

$$\frac{100.00 - 62.50}{200 - 50} = \frac{37.50}{150} = 0.25$$

You can verify that choosing any one other pair of data points will give a slope of 0.25. The data are linear.

(b) The data from Table 1.25 are plotted below.

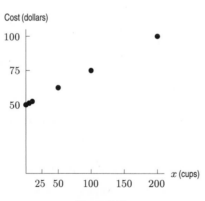

Figure 1.10

(c) Place a ruler on these points. You will see that they appear to lie on a straight line. The slope of the line equals the rate of change of the function, which is 0.25. Using units, we note that

$$\frac{\$52.50 - \$50.00}{10 \text{ cups} - 0 \text{ cups}} = \frac{\$2.50}{10 \text{ cups}} = \frac{\$0.25}{\text{cup}}.$$

In other words, the price for each additional cup of coffee is $\$0.25$.

(d) The vendor has fixed start-up costs for this venture, i.e. cart rental, insurance, salary, etc.

17. We know that the area of a circle of radius r is

$$\text{Area } = \pi r^2$$

while its circumference is given by

$$\text{Circumference } = 2\pi r.$$

Thus, a table of values for area and circumference is

Table 1.2

Radius	0	1	2	3	4	5	6
Area	0	π	4π	9π	16π	25π	36π
Circumference	0	2π	4π	6π	8π	10π	12π

(a) In the area function we see that the rate of change between pairs of points does not remain constant and thus the function is not linear. For example, the rate of change between the points $(0,0)$ and $(2,4\pi)$ is not equal to the rate of change between the points $(3,9\pi)$ and $(6,36\pi)$. The rate of change between $(0,0)$ and $(2,4\pi)$ is

$$\frac{\Delta\text{area}}{\Delta\text{radius}} = \frac{4\pi - 0}{2 - 0} = \frac{4\pi}{2} = 2\pi$$

while the rate of change between $(3,9\pi)$ and $(6,36\pi)$ is

$$\frac{\Delta\text{area}}{\Delta\text{radius}} = \frac{36\pi - 9\pi}{6 - 3} = \frac{27\pi}{3} = 9\pi.$$

On the other hand, if we take only pairs of points from the circumference function, we see that the rate of change remains constant. For instance, for the pair $(0,0)$, $(1,2\pi)$ the rate of change is

$$\frac{\Delta\text{circumference}}{\Delta\text{radius}} = \frac{2\pi - 0}{1 - 0} = \frac{2\pi}{1} = 2\pi.$$

For the pair $(2,4\pi)$, $(4,8\pi)$ the rate of change is

$$\frac{\Delta\text{circumference}}{\Delta\text{radius}} = \frac{8\pi - 4\pi}{4 - 2} = \frac{4\pi}{2} = 2\pi.$$

For the pair $(1,2\pi)$, $(6,12\pi)$ the rate of change is

$$\frac{\Delta\text{circumference}}{\Delta\text{radius}} = \frac{12\pi - 2\pi}{6 - 1} = \frac{10\pi}{5} = 2\pi.$$

Picking any pair of data points would give a rate of change of 2π.

(b) The graphs for area and circumference as indicated in Table 1.2 are shown in Figure 1.11 and Figure 1.12.

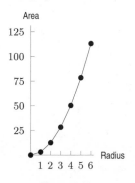

Figure 1.11

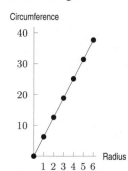

Figure 1.12

(c) From part (a) we see that the rate of change of the circumference function is 2π. This tells us that for a given circle, when we increase the length of the radius by one unit, the length of the circumference would increase by 2π units. Equivalently, if we decreased the length of the radius by one unit, the length of the circumference would decrease by 2π.

21. (a) $F = 2C + 30$

(b) Since we are finding the difference for a number of values, it would perhaps be easier to find a formula for the difference:

$$\text{Difference} = \text{Approximate value} - \text{Actual value}$$
$$= (2C + 30) - \left(\frac{9}{5}C + 32\right) = \frac{1}{5}C - 2.$$

If the Celsius temperature is $-5°$, $(1/5)C - 2 = (1/5)(-5) - 2 = -1 - 2 = -3$. This agrees with our results above.

Similarly, we see that when $C = 0$, the difference is $(1/5)(0) - 2 = -2$ or 2 degrees too low. When $C = 15$, the difference is $(1/5)(15) - 2 = 3 - 2 = 1$ or 1 degree too high. When $C = 30$, the difference is $(1/5)(30) - 2 = 6 - 2 = 4$ or 4 degrees too high.

(c) We are looking for a temperature C, for which the difference between the approximation and the actual formula is zero.

$$\frac{1}{5}C - 2 = 0$$
$$\frac{1}{5}C = 2$$
$$C = 10$$

Another way we can solve for a temperature C is to equate our approximation and the actual value.

$$\text{Approximation} = \text{Actual value}$$
$$2C + 30 = 1.8C + 32,$$
$$0.2C = 2$$
$$C = 10$$

So the approximation agrees with the actual formula at $10°$ Celsius.

25. Most functions look linear if viewed in a small enough window. This function is not linear. We see this by graphing the function in the larger window $-100 \leq x \leq 100$, $-20 \leq y \leq 20$.

29. (a) The inequalities place b to the right of a and $f(b)$ higher than $f(a)$. Since the function is linear, the graph is a line. See Figure 1.13.

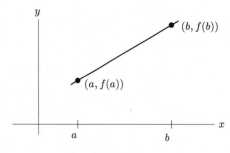

Figure 1.13

(b) The slope is

$$m = \frac{\Delta y}{\Delta x} = \frac{f(b) - f(a)}{b - a}.$$

Solutions for Section 1.4

Exercises

1. Rewriting in slope-intercept form:

$$3x + 5y = 20$$
$$5y = 20 - 3x$$
$$y = \frac{20}{5} - \frac{3x}{5}$$
$$y = 4 - \frac{3}{5}x$$

5. Rewriting in slope-intercept form:

$$5x - 3y + 2 = 0$$
$$-3y = -2 - 5x$$
$$y = \frac{-2}{-3} - \frac{5}{-3}x$$
$$y = \frac{2}{3} + \frac{5}{3}x$$

9. Rewriting in slope-intercept form:

$$\frac{x + y}{7} = 3$$
$$x + y = 21$$
$$y = 21 - x$$

13. Since we know the x-intercept and y-intercepts are $(3, 0)$ and $(0, -5)$ respectively, we can find the slope:

$$\text{slope} = m = \frac{-5 - 0}{0 - 3} = \frac{-5}{-3} = \frac{5}{3}.$$

We can then put the slope and y-intercept into the general equation for a line.

$$y = -5 + \frac{5}{3}x.$$

17. Since the function is linear, we can choose any two points to find its formula. We use the form

$$q = b + mp$$

to get the number of bottles sold as a function of the price per bottle. We use the two points $(0.50, 1500)$ and $(1.00, 500)$. We begin by finding the slope, $\Delta q / \Delta p = (500 - 1500)/(1.00 - 0.50) = -2000$. Next, we substitute a point into our equation using our slope of -2000 bottles sold per dollar increase in price and solve to find b, the q-intercept. We use the point $(1.00, 500)$:

$$500 = b - 2000 \cdot 1.00$$
$$2500 = b.$$

Therefore,

$$q = 2500 - 2000p.$$

21. Since the function is linear, we can use any two points (from the graph) to find its formula. We use the form

$$u = b + mn$$

to get the meters of shelf space used as a function of the number of different medicines stocked. We use the two points $(60, 5)$ and $(120, 10)$. We begin by finding the slope, $\Delta u / \Delta n = (10 - 5)/(120 - 60) = 1/12$. Next, we substitute a point into our equation using our slope of $1/12$ meters of shelf space per medicine and solve to find b, the u-intercept. We use the point $(60, 5)$:

$$5 = b + (1/12) \cdot 60$$
$$0 = b.$$

Therefore,

$$u = (1/12)n.$$

The fact that $b = 0$ is not surprising, since we would expect that, if no medicines are stocked, they should take up no shelf space.

Problems

25. (a) A table of the allowable combinations of sesame and poppy seed rolls is shown below.

Table 1.3

s, sesame seed rolls	0	1	2	3	4	5	6	7	8	9	10	11	12
p, poppy seed rolls	12	11	10	9	8	7	6	5	4	3	2	1	0

(b) The sum of s and p is 12. So we can write $s + p = 12$, or $p = 12 - s$.

(c)

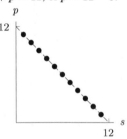

Figure 1.14

29. (a) The results are in Table 1.4.

Table 1.4

t	0	1	2	3	4
$v = f(t)$	1000	990.2	980.4	970.6	960.8

(b) The speed of the bullet is decreasing at a constant rate of 9.8 meters/sec every second. To confirm this, calculate the rate of change in velocity over every second. We get

$$\frac{\Delta v}{\Delta t} = \frac{990.2 - 1000}{1 - 0} = \frac{980.4 - 990.2}{2 - 1} = \frac{970.6 - 980.4}{3 - 2} = \frac{960.8 - 970.6}{4 - 3} = -9.8.$$

Since the value of $\Delta v / \Delta t$ comes out the same, -9.8, for every interval, we can say that the bullet is slowing down at a constant rate. This makes sense as the constant force of gravity acts to slow the upward moving bullet down at a constant rate.

(c) The slope, -9.8, is the rate at which the velocity is changing. The v-intercept of 1000 is the initial velocity of the bullet. The t-intercept of $1000/9.8 = 102.04$ is the time at which the bullet stops moving and starts to head back to Earth.

(d) Since Jupiter's gravitational field would exert a greater pull on the bullet, we would expect the bullet to slow down at a faster rate than a bullet shot from earth. On earth, the rate of change of the bullet is -9.8, meaning that the bullet is slowing down at the rate of 9.8 meters per second. On Jupiter, we expect that the coefficient of t, which represents the rate of change, to be a more negative number (less than -9.8). Similarly, since the gravitational pull near the surface of the moon is less, we expect that the bullet would slow down at a lesser rate than on earth. So, the coefficient of t should be a less negative number (greater than -9.8 but less than 0).

33.

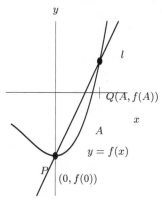

Figure 1.15

Using points $P = (0, f(0))$ and $Q = (A, f(A))$, we can find the slope of the line to be

$$m = \frac{\Delta y}{\Delta x} = \frac{f(A) - f(0)}{A - 0} = \frac{f(A) - f(0)}{A}.$$

Since the y-intercept of l is $b = f(0)$, we have

$$y = f(0) + \frac{f(A) - f(0)}{A} x.$$

37. (a) We know that $r = 1/t$. Table 1.5 gives values of r. From the table, we see that $\Delta r/\Delta H \approx 0.01/2 = 0.005$, so $r = b + 0.005H$. Solving for b, we have

$$0.070 = b + 0.005 \cdot 20$$
$$b = 0.070 - 0.1 = -0.03.$$

Thus, a formula for r is given by $r = 0.005H - 0.03$.

Table 1.5 *Development time t (in days) for an organism as a function of ambient temperature H (in °C)*

H, °C	20	22	24	26	28	30
r, rate	0.070	0.080	0.090	0.100	0.110	0.120

(b) From Problem 36, we know that if $r = b + kH$ then the number of degree-days is given by $S = 1/k$. From part (a) of this problem, we have $k = 0.005$, so $S = 1/0.005 = 200$.

Solutions for Section 1.5

Exercises

1. **(a)** Since the slopes are 2 and 3, we see that $y = -2 + 3x$ has the greater slope.
   (b) Since the y-intercepts are -1 and -2, we see that $y = -1 + 2x$ has the greater y-intercept.

5. **(a)** is (V), because slope is positive, vertical intercept is negative
   (b) is (IV), because slope is negative, vertical intercept is positive
   (c) is (I), because slope is 0, vertical intercept is positive
   (d) is (VI), because slope and vertical intercept are both negative
   (e) is (II), because slope and vertical intercept are both positive
   (f) is (III), because slope is positive, vertical intercept is 0
   (g) is (VII), because it is a vertical line with positive x-intercept.

9. These lines are parallel because they have the same slope, 5.

13. These lines are neither parallel nor perpendicular. They do not have the same slopes, nor are their slopes negative reciprocals (if they were, one of the slopes would be negative).

Problems

17. The graphs are shown in Figure 1.16.

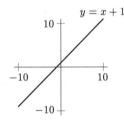
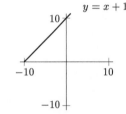
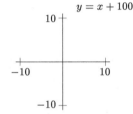

Figure 1.16

   (a) As b becomes larger, the graph moves higher and higher up, until it disappears from the viewing rectangle.
   (b) There are many correct answers, one of which is $y = x - 100$.

21. Since P is the x-intercept, we know that point P has y-coordinate $= 0$, and if the x-coordinate is x_0, we can calculate the slope of line l using $P(x_0, 0)$ and the other given point $(0, -2)$.

$$m = \frac{-2 - 0}{0 - x_0} = \frac{-2}{-x_0} = \frac{2}{x_0}.$$

   We know this equals 2, since l is parallel to $y = 2x + 1$ and therefore must have the same slope. Thus we have

$$\frac{2}{x_0} = 2.$$

   So $x_0 = 1$ and the coordinates of P are $(1, 0)$.

25. **(a)** The three formulas are linear with b being the fixed rate and m being the cost per mile. The formulas are,

$$\text{Company A} = 20 + 0.2x$$
$$\text{Company B} = 35 + 0.1x$$
$$\text{Company C} = 70.$$

(b)

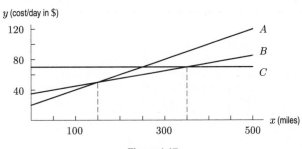

Figure 1.17

(c) The slope is the rate charged for each mile, and its units are dollars per mile. The vertical intercept is the fixed cost—what you pay for renting the car for a day, not considering mileage charges.

(d) By reading Figure 1.17 we see A is cheapest if you drive less than 150 miles; B is cheapest if you drive between 150 and 350 miles; C is cheapest if you drive more than 350 miles. We would expect A to be the cheapest for a small number of miles since it has the lowest fixed rate and C to be the cheapest for a large number of miles since it does not charge per mile.

29. (a) To have no points in common the lines will have to be parallel and distinct. To be parallel their slopes must be the same, so $m_1 = m_2$. To be distinct we need $b_1 \neq b_2$.

(b) To have all points in common the lines will have to be parallel and the same. To be parallel their slopes must be the same, so $m_1 = m_2$. To be the same we need $b_1 = b_2$.

(c) To have exactly one point in common the lines will have to be nonparallel. To be nonparallel their slopes must be distinct, so $m_1 \neq m_2$.

(d) It is not possible for two lines to meet in just two points.

Solutions for Section 1.6

1. (a) The points are graphed in Figure 1.18.

(b) See Figure 1.18.

(c) Since the points all seem to lie on a line, the correlation coefficient is close to one.

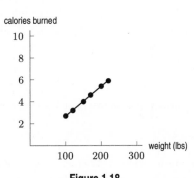

Figure 1.18

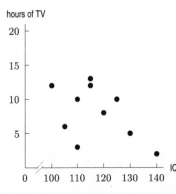

Figure 1.19

5. (a) See Figure 1.19.

(b) The scatterplot suggests that as IQ increases, the number of hours of TV viewing decreases. The points, though, are not close to being on a line, so a reasonable guess is $r \approx -0.5$.

(c) A calculator gives the regression equation $y = 27.5139 - 0.1674x$ with $r = -0.5389$.

Solutions for Chapter 1 Review

Exercises

1. At the two points where the graph breaks (marked A and B in Figure 1.20), there are two y values for a single x value. The graph does not pass the vertical line test. Thus, y is not a function of x.

    Similarly, x is not a function of y because there are many y values that give two x values (For example, $y = 0$.)

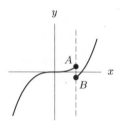

Figure 1.20

5. Both of the relationships are functions because any quantity of gas determines the quantity of coffee that can be bought, and vice versa. For example, if you buy 30 gallons of gas, spending $60, you buy 4 pounds of coffee.

9. This table could not represent a linear function, because the rate of change of $q(\lambda)$ is not constant. Consider the first three points in the table. Between $\lambda = 1$ and $\lambda = 2$, we have $\Delta\lambda = 1$ and $\Delta q(\lambda) = 2$, so the rate of change is $\Delta q(\lambda)/\Delta\lambda = 2$. Between $\lambda = 2$ and $\lambda = 3$, we have $\Delta\lambda = 1$ and $\Delta q(\lambda) = 4$, so the rate of change is $\Delta q(\lambda)/\Delta\lambda = 4$. Thus, the function could not be linear.

13. **(a)** Since the slopes are -2 and -3, we see that $y = 5 - 2x$ has the greater slope.
    (b) Since the y-intercepts are 5 and 7, we see that $y = 7 - 3x$ has the greater y-intercept.

17. These lines are perpendicular because one slope, $-\frac{1}{14}$, is the negative reciprocal of the other, 14.

21. The line $y + 4x = 7$ has slope -4. Therefore the parallel line has slope -4 and equation $y - 5 = -4(x - 1)$ or $y = -4x + 9$. The perpendicular line has slope $\frac{-1}{(-4)} = \frac{1}{4}$ and equation $y - 5 = \frac{1}{4}(x - 1)$ or $y = 0.25x + 4.75$.

Problems

25. **(a)** We have $r_m(0) - r_h(0) = 216 - 31 = 185$. This tells us that in 1990 (year $t = 0$), the name Hannah was ranked 185 places higher than Madison on the list of most popular names. (Recall that the lower the ranking, the higher a names position on the list.)
    (b) We have $r_m(11) - r_h(11) = 2 - 3 = -1$. This tells us that in 2001 (year $t = 11$), the name Hannah was ranked 1 place lower than Madison on the list of most popular names.
    (c) We have $r_m(t) < r_a(t)$ for $t = 10$ and $t = 11$. This tells us that the name Madison was ranked higher than the name Alexis on the list of most popular names in the years 2000 and 2001.

29. In Figure 1.21 the graph of the hair length is steepest just after each haircut, assumed to be at the beginning of each year. As the year progresses, the growth is slowed by split ends. By the end of the year, the hair is breaking off as fast as it is growing, so the graph has leveled off. At this time the hair is cut again. Once again it grows until slowed by the split ends. Then it is cut. This continues for five years when the longest hairs fall out because they have come to the end of their natural lifespan.

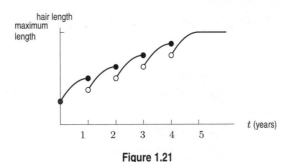

Figure 1.21

33. **(a)** At 40 mph, fuel consumption is about 28 mpg, so the fuel used is $300/28 = 10.71$ gallons.

   (b) At 60 mph, fuel consumption is about 29 mpg. At 70 mph, fuel consumption is about 28 mpg. Therefore, on a 200 mile trip

$$\text{Fuel saved} = \frac{200}{28} - \frac{200}{29} = 0.25 \text{ gallons.}$$

   (c) The most fuel-efficient speed is where mpg is a maximum, which is about 55 mph.

37. **(a)** Adding the male total to the female total gives $x + y$, the total number of applicants.

   (b) Of the men who apply, 15% are accepted. So $0.15x$ male applicants are accepted. Likewise, 18% of the women are accepted so we have $0.18y$ women accepted. Summing the two tells us that $0.15x + 0.18y$ applicants are accepted.

   (c) The number accepted divided by the number who applied times 100 gives the percentage accepted. This expression is

$$\frac{(0.15)x + (0.18)y}{x + y}(100), \quad \text{or} \quad \frac{15x + 18y}{x + y}.$$

41. **(a)** $C(175) = 11{,}375$, which means that it costs \$11,375 to produce 175 units of the good.

   (b) $C(175) - C(150) = 125$, which means that the cost of producing 175 units is \$125 greater than the cost of producing 150 units. That is, the cost of producing the additional 25 units is an additional \$125.

   (c) $\dfrac{C(175) - C(150)}{175 - 150} = \dfrac{125}{25} = 5$, which means that the average per-unit cost of increasing production to 175 units from 150 units is \$5.

45. **(a)** Since i is linear, we can write

$$i(x) = b + mx.$$

Since $i(10) = 25$ and $i(20) = 50$, we have

$$m = \frac{50 - 25}{20 - 10} = 2.5.$$

So,

$$i(x) = b + 2.5x.$$

Using $i(10) = 25$, we can solve for b:

$$i(10) = b + 2.5(10)$$
$$25 = b + 25$$
$$b = 0.$$

Our formula then is

$$i(x) = 2.5x.$$

   (b) The increase in risk associated with *not* smoking is $i(0)$. Since there is no increase in risk for a non-smoker, we have $i(0) = 0$.

   (c) The slope of $i(x)$ tells us that the risk increases by a factor of 2.5 with each additional cigarette a person smokes per day.

49. (a) See Figure 1.22.

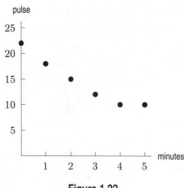

Figure 1.22

(b) For $0 \le t \le 4$, the pulse values nearly lie on a straight line.

(c) The correlation is close to $r = -1$ for time less than 4 minutes. After 4 minutes, the pulse rate reaches its normal, constant level, and there would be no correlation.

53. In Figure 1.23 the decision to spend all c dollars of your money on apples is represented by the x-intercept; the decision to spend it all on bananas is represented the y-intercept. If we decide to spend all of our money on either all apples or all bananas, and bananas are cheaper, then we would be able to purchase more bananas than apples for our c dollars. So, we want the line for which the y-intercept is greater than the x-intercept. If we look at line l_1 we see that the y-intercept is greater than 10 and the x–intercept is less than 10. Thus, l_1 represents the case where we can buy more bananas than apples, so apples must be more expensive than bananas.

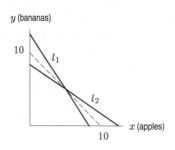

Figure 1.23

CHECK YOUR UNDERSTANDING

1. False. $f(t)$ is functional notation, meaning that f is a function of the variable t.

5. True. The number of people who enter a store in a day and the total sales for the day are related, but neither quantity is uniquely determined by the other.

9. True. A circle does not pass the vertical line test.

13. True. This is the definition of an increasing function.

17. False. Parentheses must be inserted. The correct ratio is $\dfrac{(10 - 2^2) - (10 - 1^2)}{2 - 1} = -3$.

21. False. Writing the equation as $y = (-3/2)x + 7/2$ shows that the slope is $-3/2$.

25. True. A constant function has slope zero. Its graph is a horizontal line.

29. True. At $y = 0$, we have $4x = 52$, so $x = 13$. The x-intercept is $(13, 0)$.

33. False. Substitute the point's coordinates in the equation: $-3 - 4 \neq -2(4 + 3)$.

37. False. The first line does but the second, in slope-intercept form, is $y = (1/8)x + (1/2)$, so it crosses the y-axis at $y = 1/2$.

41. True. The point $(1, 3)$ is on both lines because $3 = -2 \cdot 1 + 5$ and $3 = 6 \cdot 1 - 3$.

45. True. The slope, $\Delta y / \Delta x$ is undefined because Δx is zero for any two points on a vertical line.

49. False. For example, in children there is a high correlation between height and reading ability, but it is clear that neither causes the other.

53. True. There is a perfect fit of the line to the data.

Solutions to Tools for Chapter 1

1.
$$3x = 15$$
$$\frac{3x}{3} = \frac{15}{3}$$
$$x = 5$$

5.
$$y - 5 = 21$$
$$y = 26$$

9.
$$13t + 2 = 47$$
$$13t = 45$$
$$\frac{13t}{13} = \frac{45}{13}$$
$$t = \frac{45}{13}$$

13. We first distribute $\frac{5}{3}(y + 2)$ to obtain:

$$\frac{5}{3}(y + 2) = \frac{1}{2} - y$$
$$\frac{5}{3}y + \frac{10}{3} = \frac{1}{2} - y$$
$$\frac{5}{3}y + y = \frac{1}{2} - \frac{10}{3}$$
$$\frac{5}{3}y + \frac{3y}{3} = \frac{3}{6} - \frac{20}{6}$$
$$\frac{8y}{3} = -\frac{17}{16}$$
$$\left(\frac{3}{8}\right)\frac{8y}{3} = \left(\frac{3}{8}\right)\left(-\frac{17}{6}\right)$$
$$y = -\frac{17}{16}.$$

17. Expanding yields

$$1.06s - 0.01(248.4 - s) = 22.67s$$
$$1.06s - 2.484 + 0.01s = 22.67s$$
$$-21.6s = 2.484$$
$$s = -0.115.$$

21. We have

$$C = \frac{5}{9}(F - 32)$$
$$\frac{9C}{5} = F - 32$$
$$F = \frac{9}{5}C + 32$$

25. We collect all terms involving x and then divide by $2a$:

$$ab + ax = c - ax$$
$$2ax = c - ab$$
$$x = \frac{c - ab}{2a}.$$

29. Solving for x:

$$\frac{a - cx}{b + dx} + a = 0$$
$$\frac{a - cx}{b + dx} = -a$$
$$a - cx = -a(b + dx) = -ab - adx$$
$$adx - cx = -ab - a$$
$$(ad - c)x = -a(b + 1)$$
$$x = -\frac{a(b + 1)}{ad - c}.$$

33. Adding the two equations to eliminate y, we have

$$2x = 8$$
$$x = 4.$$

Using $x = 4$ in the first equation gives

$$4 + y = 3,$$

so

$$y = -1.$$

37. We substitute the expression $-\frac{3}{5}x + 6$ for y in the first equation.

$$2x + 3y = 7$$
$$2x + 3\left(-\frac{3}{5}x + 6\right) = 7$$
$$2x - \frac{9}{5}x + 18 = 7 \quad \text{or}$$
$$\frac{10}{5}x - \frac{9}{5}x + 18 = 7$$
$$\frac{1}{5}x + 18 = 7$$
$$\frac{1}{5}x = -11$$
$$x = -55$$
$$y = -\frac{3}{5}(-55) + 6$$
$$y = 39$$

41. We set the equations $y = x$ and $y = 3 - x$ equal to one another.

$$x = 3 - x$$
$$2x = 3$$
$$x = \frac{3}{2} \quad \text{and} \quad y = \frac{3}{2}$$

So the point of intersection is $x = 3/2, y = 3/2$.

CHAPTER TWO

Solutions for Section 2.1

Exercises

1. To evaluate when $x = -7$, we substitute -7 for x in the function, giving $f(-7) = -\dfrac{7}{2} - 1 = -\dfrac{9}{2}$.

5. To solve for x when $f(x) = 8$, we substitute 8 for $f(x)$ in the function and solve for x:

$$8 = x^2 - 8$$
$$16 = x^2$$
$$\pm\sqrt{16} = x$$
$$\pm 4 = x.$$

9. (a) Substituting $x = 0$ gives $g(0) = 0^2 - 5(0) + 6 = 6$.
  (b) Setting $g(x) = 0$ and solving gives $x^2 - 5x + 6 = 0$.
   Factoring gives $(x - 2)(x - 3) = 0$, so $x = 2, 3$.

Problems

13. (a) The table shows $f(6) = 3.7$, so $t = 6$. In a typical June, Chicago has 3.7 inches of rain.
  (b) First evaluate $f(2) = 1.8$. Solving $f(t) = 1.8$ gives $t = 1$ or $t = 2$. Chicago has 1.8 inches of rain in January and in February.

17. (a) In order to find $f(0)$, we need to find the value which corresponds to $x = 0$. The point $(0, 24)$ seems to lie on the graph, so $f(0) = 24$.
  (b) Since $(1, 10)$ seems to lie on this graph, we can say that $f(1) = 10$.
  (c) The point that corresponds to $x = b$ seems to be about $(b, -7)$, so $f(b) = -7$.
  (d) When $x = c$, we see that $y = 0$, so $f(c) = 0$.
  (e) When your input is d, the output is about 20, so $f(d) = 20$.

21. (a) We calculate the values of $f(x)$ and $g(x)$ using the formulas given in Table 2.1.

Table 2.1

x	-2	-1	0	1	2
$f(x)$	6	2	0	0	2
$g(x)$	6	2	0	0	2

The pattern is that $f(x) = g(x)$ for $x = -2, -1, 0, 1, 2$. Based on this, we might speculate that f and g are really the same function. This is, in fact, the case, as can be verified algebraically:

$$f(x) = 2x(x - 3) - x(x - 5)$$
$$= 2x^2 - 6x - x^2 + 5x$$
$$= x^2 - x$$
$$= g(x).$$

Their graphs are the same, and are shown in Figure 2.1.

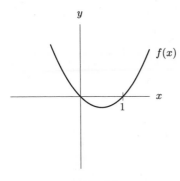

Figure 2.1

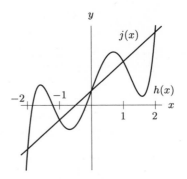

Figure 2.2

(b) Using the formulas for $h(x)$ and $j(x)$, we obtain Table 2.2.

Table 2.2

x	-2	-1	0	1	2
$h(x)$	-3	-1	1	3	5
$j(x)$	-3	-1	1	3	5

The pattern is that $h(x) = j(x)$ for $x = -2, -1, 0, 1, 2$. Based on this, we might speculate that h and j are really the same function. The graphs of these functions are shown in Figure 2.2. We see that the graphs share only the points of the table and are thus two different functions.

25. **(a)** Her tax is $4056 on the first $65,000 plus 6.85% of the remaining $3000:

$$\text{Tax owed} = \$4056 + 0.0685(\$3000) = \$4056 + \$205.50 = \$4261.50.$$

(b) Her taxable income, $T(x)$, is 80% of her total income, or 80% of x. So $T(x) = 0.8x$.

(c) Her tax owed is $4056 plus 6.85% of her taxable income over $65,000. Since her taxable income is $0.8x$, her taxable income over $65,000 is $0.8x - 65,000$. Therefore,

$$L(x) = 4056 + 0.0685(0.8x - 65000),$$

so multiplying out and simplifying, we obtain

$$L(x) = 0.0548x - 396.5.$$

(d) Evaluating for $x = \$85,000$, we have

$$L(85,000) = 4056 + 0.0685(0.8(85,000) - 65,000)$$
$$= \$4261.50.$$

The values are the same.

Solutions for Section 2.2

Exercises

1. The graph of $f(x) = 1/x^2$ for $-1 \le x \le 1$ is shown in Figure 2.3. From the graph, we see that $f(x) = 1$ at $x = -1$ and $x = 1$. As we approach 0 from 1 or from -1, the graph increases without bound. The lower limit of the range is 1, while there is no upper limit. Thus, on the domain $-1 \le x \le 1$, the range is $f(x) \ge 1$.

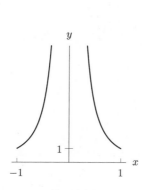

Figure 2.3

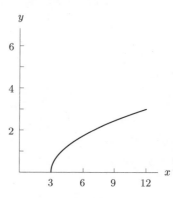

Figure 2.4

5. The graph of $f(x) = \sqrt{x - 3}$ is given in Figure 2.4.   The domain is all real $x \ge 3$; the range is all $f(x) \ge 0$.

9. The graph of $f(x) = x^2 - 4$ is given in Figure 2.5.

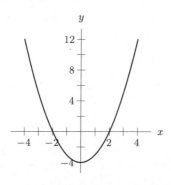

Figure 2.5

The domain is all real x; the range is all $f(x) \ge -4$.

13. Since for any value of x that you might choose you can find a corresponding value of $m(x)$, we can say that the domain of $m(x) = 9 - x$ is all real numbers.
    For any value of $m(x)$ there is a corresponding value of x. So the range is also all real numbers.

17. We can take the cube root of any number, so the domain is all real numbers. By using appropriate input values we can get any real number as a result, so the range is all real numbers.

21. Any number can be squared, so the domain is all real numbers. Since $(x - 3)^2$ is always greater than or equal to zero, we see that $f(x) = (x - 3)^2 + 2 \ge 2$. Thus, the range is all real numbers ≥ 2.

25. The domain is $2 \le x \le 6$. The range is $1 \le f(x) \le 3$.

Problems

29. (a) From the table we find that a 200 lb person uses 5.4 calories per minute while walking. So a half-hour, or a 30 minute, walk burns $30(5.4) = 162$ calories.

(b) The number of calories used per minute is approximately proportional to the person's weight. The relationship is an approximately linear increasing function, where weight is the independent variable and number of calories burned is the dependent variable.

(c) (i) Since the function is approximately linear, its equation is $c = b + mw$, where c is the number of calories and w is weight. Using the first two values in the table, the slope is

$$m = \frac{3.2 - 2.7}{120 - 100} = \frac{0.5}{20} = 0.025 \text{ cal/lb.}$$

Using the point $(100, 2.7)$ we have

$$2.7 = b + 0.025(100)$$
$$b = 0.2.$$

So the equation is $c = 0.2 + 0.025w$. See Figure 2.6. All the values given lie on this line with the exception of the last two which are slightly above it.

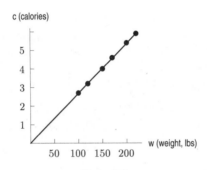

Figure 2.6

(ii) The intercept $(0, 0.2)$ is the number of calories burned by a weightless runner. Since 0.2 is a small number, most of the calories burned appear to be due to moving a person's weight. Other methods of finding the equation of the the line may give other values for the vertical intercept, but all values are close to 0.

(iii) Domain $0 < w$; range $0 < c$

(iv) Evaluating the function at $w = 135$,

$$\text{Calories} = 0.2 + 0.025(135) \approx 3.6.$$

33. (a) Substituting $t = 0$ into the formula for $p(t)$ shows that $p(0) = 50$, meaning that there were 50 rabbits initially. Using a calculator, we see that $p(10) \approx 131$, which tells us there were about 131 rabbits after 10 months. Similarly, $p(50) \approx 911$ means there were about 911 rabbits after 50 months.

(b) The graph in Figure 2.7 tells us that the rabbit population grew quickly at first but then leveled off at about 1000 rabbits after around 75 months or so. It appears that the rabbit population increased until it reached the island's capacity.

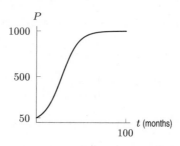

Figure 2.7

(c) From the graph in Figure 2.7, we see that the range is $50 \leq p(t) \leq 1000$. This tells us that (for $t \geq 0$) the number of rabbits is no less than 50 and no more than 1000.

(d) The smallest population occurred when $t = 0$. At that time, there were 50 rabbits. As t gets larger and larger, $(0.9)^t$ gets closer and closer to 0. Thus, as t increases, the denominator of

$$p(t) = \frac{1000}{1 + 19(0.9)^t}$$

decreases. As t increases, the denominator $1 + 19(0.9)^t$ gets close to 1 (try $t = 100$, for example). As the denominator gets closer to 1, the fraction gets closer to 1000. Thus, as t gets larger and larger, the population gets closer and closer to 1000. Thus, the range is $50 \leq p(t) < 1000$.

Solutions for Section 2.3

Exercises

1. $f(x) = \begin{cases} -1, & -1 \leq x < 0 \\ 0, & 0 \leq x < 1 \\ 1, & 1 \leq x < 2 \end{cases}$ is shown in Figure 2.8.

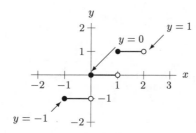

Figure 2.8

5. We find the formulas for each of the lines. For the first, we use the two points we have, $(1, 4)$ and $(3, 2)$. We find the slope: $(2 - 4)/(3 - 1) = -1$. Using the slope of -1, we solve for the y-intercept:

$$4 = b - 1 \cdot 1$$
$$5 = b.$$

Thus, the first line is $y = 5 - x$, and it is for the part of the function where $x < 3$. Notice that we do not use this formula for the value $x = 3$.

We follow the same method for the second line, using the points $(3, \frac{1}{2})$ and $(5, \frac{3}{2})$. We find the slope: $(\frac{3}{2} - \frac{1}{2})/(5 - 3) = \frac{1}{2}$. Using the slope of $\frac{1}{2}$, we solve for the y-intercept:

$$\frac{1}{2} = b + \frac{1}{2} \cdot 3$$
$$-1 = b.$$

Thus, the second line is $y = -1 + \frac{1}{2}x$, and it is for the part of the function where $x \geq 3$.
Therefore, the function is:

$$y = \begin{cases} 5 - x & \text{for} & x < 3 \\ -1 + \frac{1}{2}x & \text{for} & x \geq 3. \end{cases}$$

Problems

9. **(a)** Yes, because every value of x is associated with exactly one value of y.
   (b) No, because some values of y are associated with more than one value of x.
   (c) $y = 1, 2, 3, 4$.

13. **(a)** The depth of the driveway is 1 foot or 1/3 of a yard. The volume of the driveway is the product of the three dimensions, length, width and depth. So,

$$\begin{array}{c} \text{Volume of} \\ \text{gravel needed} \end{array} = \text{Length} \cdot \text{Width} \cdot \text{Depth} = (L)(6)(1/3) = 2L.$$

Since he buys 10 cubic yards more than needed,

$$n(L) = 2L + 10.$$

   (b) The length of a driveway is not less than 5 yards, so the domain of n is all real numbers greater than or equal to 5. The contractor can buy only 1 cubic yd at a time so the range only contains integers. The smallest value of the range occurs for the shortest driveway, when $L = 5$. If $L = 5$, then $n(5) = 2(5) + 10 = 20$. Although very long driveways are highly unlikely, there is no upper limit on L, so no upper limit on $n(L)$. Therefore the range is allintegers greater than or equal to 20. See Figure 2.9.

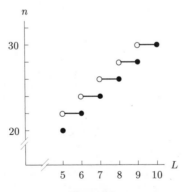

Figure 2.9

   (c) If $n(L) = 2L + 10$ was not intended to represent a quantity of gravel, then the domain and range of n would be all real numbers.

17. **(a)** Each signature printed costs $0.14, and in a book of p pages, there are at least $p/16$ signatures. In a book of 128 pages, there are

$$\frac{128}{16} = 8 \text{ signatures,}$$

$$\text{Cost for 128 pages } = 0.14(8) = \$1.12.$$

A book of 129 pages requires 9 signatures, although the ninth signature is used to print only 1 page. Therefore,

$$\text{Cost for 129 pages } = \$0.14(9) = \$1.26.$$

To find the cost of p pages, we first find the number of signatures. If p is divisible by 16, then the number of signatures is $p/16$ and the cost is

$$C(p) = 0.14 \left(\frac{p}{16} \right).$$

If p is not divisible by 16, the number of signatures is $p/16$ rounded up to the next highest integer and the cost is 0.14 times that number. In this case, it is hard to write a formula for $C(p)$ without a symbol for "rounding up."

(b) The number of pages, p, is greater than zero. Although it is possible to have a page which is only half filled, we do not say that a book has 124 1/2 pages, so p must be an integer. Therefore, the domain of $C(p)$ is $p > 0$, p an integer. Because the cost of a book increases by multiples of $0.14 (the cost of one signature), the range of $C(p)$ is $C > 0$, C an integer multiple of $0.14,

(c) For 1 to 16 pages, the cost is $0.14, because only 1 signature is required. For 17 to 32 pages, the cost is $0.28, because 2 signatures are required. These data are continued in Table 2.3 for $0 \le p \le 128$, and they are plotted in Figure 2.10. A closed circle represents a point included on the graph, and an open circle indicates a point excluded from the graph. The unbroken lines in Figure 2.10 suggest, erroneously, that *fractions* of pages can be printed. It would be more accurate to draw each step as 16 separate dots instead of as an unbroken line.

Table 2.3 *The cost C for printing a book of p pages*

p, pages	C, dollars
1-16	0.14
17-32	0.28
33-48	0.42
49-64	0.56
65-80	0.70
81-96	0.84
97-112	0.98
113-128	1.12

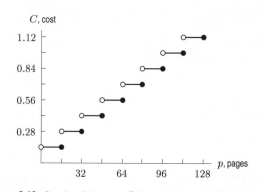

Figure 2.10: Graph of the cost C for printing a book of p pages

Solutions for Section 2.4

Exercises

1. **(a)** Since the vertical intercept of the graph of f is $(0, 2)$, we have $f(0) = 2$.
   (b) Since the horizontal intercept of the graph of f is $(-3, 0)$, we have $f(-3) = 0$.
   (c) The function f^{-1} goes from y-values to x-values, so to evaluate $f^{-1}(0)$, we want the x-value corresponding to $y = 0$. This is $x = -3$, so $f^{-1}(0) = -3$.
   (d) Solving $f^{-1}(?) = 0$ means finding the y-value corresponding to $x = 0$. This is $y = 2$, so $f^{-1}(2) = 0$.

5. (a) $f^{-1}(3) = 0$, since $f(0) = 3$.
   (b) $f^{-1}(5) = 3$, since $f(3) = 5$.
   (c) $f^{-1}(0) = 7$, since $f(7) = 0$.

Problems

9. The inverse function, $f^{-1}(P)$, gives the year in which population is P million. Units of $f^{-1}(P)$ are years.

13. The inverse function, $f^{-1}(I)$, gives the interest rate that gives $\$I$ in interest. Units of $f^{-1}(I)$ is percent per year.

17. (a) $f(10) = 100 + 0.2 \cdot 10 = 102$ thousand dollars, the cost of producing 10 kg of the chemical.
   (b) $f^{-1}(200)$ is the quantity of the chemical which can be produced for 200 thousand dollars. Since

$$200 = 100 + 0.2q$$
$$0.2q = 100$$
$$q = \frac{100}{0.2} = 500 \text{ kg},$$

we have $f^{-1}(200) = 500$.
   (c) To find $f^{-1}(C)$, solve for q:

$$C = 100 + 0.2q$$
$$0.2q = C - 100$$
$$q = \frac{C}{0.2} - \frac{100}{0.2} = 5C - 500$$
$$f^{-1}(C) = 5C - 500.$$

21. (a) This is the fare for a ride of 3.5 miles. $C(3.5) \approx \$6.25$.
   (b) This is the number of miles you can travel for $\$3.50$. Between 1 and 2 miles the increase in cost is $\$1.50$. Setting up a proportion we have:

$$\frac{1 \text{ additional mile}}{\$1.50 \text{ additional fare}} = \frac{x \text{ additional miles}}{\$3.50 - \$2.50 \text{ additional fare}}$$

and $x = 0.67$ miles. Therefore

$$C^{-1}(\$3.5) \approx 1.67.$$

Solutions for Section 2.5

Exercises

1. The rate of change between $t = 0.2$ and $t = 0.4$ is

$$\frac{\Delta p(t)}{\Delta t} = \frac{-2.32 - (-3.19)}{0.4 - 0.2} = 4.35.$$

Similarly, we have

$$\frac{\Delta p(t)}{\Delta t} = \frac{-1.50 - (-2.32)}{0.6 - 0.4} = 4.10$$

$$\frac{\Delta p(t)}{\Delta t} = \frac{-0.74 - (-1.50)}{0.8 - 0.6} = 3.80.$$

Thus, the rate of change is decreasing, so we expect the graph to be concave down.

5. The graph appears to be concave up, as its slope becomes less negative as x increases.

9. The slope of $y = x^3$ is always increasing on the interval $x > 0$, so its graph is concave up. See Figure 2.11.

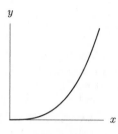

Figure 2.11

Problems

13. The function is increasing throughout. At first, the graph is concave up. As more and more people hear the rumor, the rumor spreads more slowly, which means that the graph is then concave down.

17. (a) This is a case in which the rate of decrease is constant, i.e., the change in y divided by the change in x is always the same. We see this in Table (F), where y decreases by 80 units for every decrease of 1 unit in x, and graphically in Graph (IV).

(b) Here, the change in y gets smaller and smaller relative to corresponding changes in x. In Table (G), y decreases by 216 units for a change of 1 unit in x initially, but only decreases by 6 units when x changes by 1 unit from 4 to 5. This is seen in Graph (I), where y is falling rapidly at first, but much more slowly for longer values of x.

(c) If y is the distance from the ground, we see in Table (E) that initially it is changing very slowly; by the end, however, the distance from the ground is changing rapidly. This is shown in Graph (II), where the decrease in y is larger and larger as x gets bigger.

(d) Here, y is decreasing quickly at first, then decreases only slightly for a while, then decreases rapidly again. This occurs in Table (H), where y decreases from 147 units, then 39, and finally by another 147 units. This corresponds to Graph (III).

Solutions for Section 2.6

Exercises

1. We solve for r in the equation by factoring

$$2r^2 - 6r - 36 = 0$$
$$2(r^2 - 3r - 18) = 0$$
$$2(r - 6)(r + 3) = 0.$$

The solutions are $r = 6$ and $r = -3$.

5. To find the zeros, we solve the equation

$$0 = 7x^2 + 16x + 4.$$

We see that this is factorable, as follows:

$$0 = (7x + 2)(x + 2).$$

Therefore, the zeros occur where $x = -\frac{2}{7}$ and $x = -2$.

9. The function $f(x) = (x - 1)(x - 2)$ has zeros $x = 1$ and $x = 2$. To get another function with the same zeros, we can multiply $f(x)$ by any constant: for example, let $g(x) = -7(x - 1)(x - 2)$. In general, any function of the form $y = a(x - 1)9x - 2)$ will do.

Problems

13. We solve the equation $f(t) = -16t^2 + 64t + 3 = 0$ using the quadratic formula

$$-16t^2 + 64t + 3 = 0$$

$$t = \frac{-64 \pm \sqrt{64^2 - 4(-16)3}}{2(-16)}.$$

Evaluating gives $t = -0.046$ sec and $t = 4.046$ sec; the value $t = 4.046$ sec is the time we want. The baseball hits the ground 4.046 sec after it was hit.

17. To show that the data in the table is approximated by the formula $p(x) = -0.8x^2 + 8.8x + 7.2$, we substitute $x = 0, 1, 2, 3, 4$ (for years 1992-1996) into the formula:

$$p(0) = 7.2, \ p(1) = 15.2, \ p(2) = 21.6, \ p(3) = 26.4, \ p(4) = 29.6.$$

Our results approximate the table. In the year 2004, $x = 12$, and $p(12) = -2.4$, so the model predicts -2.4% of schools will have videodisc players in 2004. This is a reasonable model for the period 1992 to 1996, but not for the year 2004 since -2.4% does not make sense. Since the x^2 term in $p(x)$ has a negative coefficient, as x increases beyond 12, the values of $p(x)$ become more negative, and so are not reasonable predictions for the percentage of schools with a videodisc player. Thus, $p(x)$ is not a good model for predicting the future.

21. **(a)** According to the figure in the text, the package was dropped from a height of 5 km.
    (b) When the package hits the ground, $h = 0$ and $d = 4430$. So, the package has moved 4430 meters forward when it lands.
    (c) Since the maximum is at $d = 0$, the parabola is of the form $h = ad^2 + b$. Since $h = 5$ at $d = 0$, $5 = a(0)^2 + b = b$, so $b = 5$. We now know that $h = ad^2 + 5$. Since $h = 0$ when $d = 4430$, we have $0 = a(4430)^2 + 5$, giving $a = \frac{-5}{(4430)^2} \approx -0.000000255$. So $h \approx -0.000000255d^2 + 5$.

Solutions for Chapter 2 Review

Exercises

1. To evaluate $p(7)$, we substitute 7 for each r in the formula:

$$p(7) = 7^2 + 5 = 54.$$

5. **(a)** $h(1) = (1)^2 + b(1)^2 + c = b + c + 1$
   (b) Substituting $b + 1$ for x in the formula for $h(x)$:

$$h(b + 1) = (b + 1)^2 + b(b + 1) + c$$
$$= (b^2 + 2b + 1) + b^2 + b + c$$
$$= 2b^2 + 3b + c + 1$$

9. (a) A table of values for $y = 1/x^2$ follows:

x	-5	-4	-3	-2	-1	$-\frac{3}{4}$	$-\frac{1}{2}$	$-\frac{1}{4}$
y	$\frac{1}{25}$	$\frac{1}{16}$	$\frac{1}{9}$	$\frac{1}{4}$	1	$\frac{16}{9}$	4	16

x	0	$\frac{1}{4}$	$\frac{1}{2}$	$\frac{3}{4}$	1	2	3	4	5
y	undefined	16	4	$\frac{16}{9}$	1	$\frac{1}{4}$	$\frac{1}{9}$	$\frac{1}{16}$	$\frac{1}{25}$

(b) The y-values in the table range from near 0 to 16, but values of x close to 0 give y-values of very large magnitude.

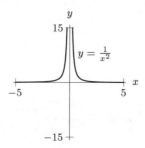

Figure 2.12

(c) Domain is all real numbers except 0.
Range is $0 < y < \infty$.
(d) Increasing: $-\infty < x < 0$.
Decreasing: $0 < x < \infty$.

13. The square root of a negative number is undefined, and so x must not be less than 4, but it can have any value greater than or equal to 4. Since $f(4) = 0$, and $f(x)$ increases as x increases, $f(x)$ is greater than or equal to zero. Thus, the domain of $f(x)$ is $x \geq 4$, and the range is $f(x) \geq 0$.

17. (a) $-3g(x) = -3(x^2 + x)$.
(b) $g(1) - x = (1^2 + 1) - x = 2 - x$.
(c) $g(x) + \pi = (x^2 + x) + \pi = x^2 + x + \pi$.
(d) $\sqrt{g(x)} = \sqrt{x^2 + x}$.
(e) $g(1)/(x + 1) = (1^2 + 1)/(x + 1) = 2/(x + 1)$.
(f) $(g(x))^2 = (x^2 + x)^2$.

Problems

21. $f\left(\dfrac{1}{1-a}\right) = \dfrac{a\dfrac{1}{1-a}}{a + \dfrac{1}{1-a}} = \dfrac{\dfrac{a}{1-a}}{\dfrac{a(1-a)+1}{1-a}} = \dfrac{a}{1-a} \cdot \dfrac{1-a}{a-a^2+1} = \dfrac{a}{a-a^2+1}.$

25. Graph of $g(x)$ is above graph of $f(x)$ for x to the right of 2. See Figure 2.13.

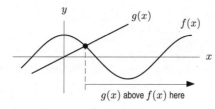

Figure 2.13

29. (a) $j(h(4)) = h^{-1}(h(4)) = 4$

(b) We don't know $j(4)$

(c) $h(j(4)) = h(h^{-1}(4)) = 4$

(d) $j(2) = 4$

(e) We don't know $h^{-1}(-3)$

(f) $j^{-1}(-3) = 5$, since $j(5) = -3$

(g) We don't know $h(5)$

(h) $(h(-3))^{-1} = (j^{-1}(-3))^{-1} = 5^{-1} = 1/5$

(i) We don't know $(h(2))^{-1}$

33. (a) $t(400) = 272$.

(b) (i) It takes 136 seconds to melt 1 gram of the compound at a temperature of $800°$C.

(ii) It takes 68 seconds to melt 1 gram of the compound at a temperature of $1600°$C.

(c) This means that $t(2x) = t(x)/2$, because if x is a temperature and $t(x)$ is a melting time, then $2x$ would be double this temperature and $t(x)/2$ would be half this melting time.

CHECK YOUR UNDERSTANDING

1. False. $f(2) = 3 \cdot 2^2 - 4 = 8$.

5. False. $W = (8 + 4)/(8 - 4) = 3$.

9. True. A fraction can only be zero if the numerator is zero.

13. False. The domain consists of all real numbers x, $x \neq 3$

17. True. Since f is an increasing function, the domain endpoints determine the range endpoints. We have $f(15) = 12$ and $f(20) = 14$.

21. True. $|x| = |-x|$ for all x.

25. True. If $x < 0$, then $f(x) = x < 0$, so $f(x) \neq 4$. If $x > 4$, then $f(x) = -x < 0$, so $f(x) \neq 4$. If $0 \leq x \leq 4$, then $f(x) = x^2 = 4$ only for $x = 2$. The only solution for the equation $f(x) = 4$ is $x = 2$.

29. False. Check to see if $f(0) = 8$, which it does not.

33. False. The output units of a function are the same as the input units of its inverse.

37. False. A straight line is neither concave up nor concave down.

41. False. The zeros are -1 and -2.

45. True. Solving $f(x) = 0$ gives the zeros of $f(x)$.

Solutions to Tools for Chapter 2

1. $3(x + 2) = 3x + 6$

5. $12(x + y) = 12x + 12y$

9. $-10r(5r + 6rs) = -50r^2 - 60r^2s$

13. $(x - 2)(x + 6) = x^2 + 6x - 2x - 12 = x^2 + 4x - 12$

17. $(12y - 5)(8w + 7) = 96wy + 84y - 40w - 35$

21. First we multiply 4 by the terms $3x$ and $-2x^2$, and expand $(5 + 4x)(3x - 4)$. Therefore,

$$\left(3x - 2x^2\right)(4) + (5 + 4x)(3x - 4) = 12x - 8x^2 + 15x - 20 + 12x^2 - 16x$$

$$= 4x^2 + 11x - 20.$$

25. The order of operations tells us to expand $(x-3)^2$ first and then multiply the result by 4. Therefore,

$$
\begin{aligned}
4(x-3)^2 + 7 &= 4(x-3)(x-3) + 7 \\
&= 4(x^2 - 3x - 3x + 9) + 7 = 4(x^2 - 6x + 9) + 7 \\
&= 4x^2 - 24x + 36 + 7 = 4x^2 - 24x + 43.
\end{aligned}
$$

29. $3y + 15 = 3(y + 5)$

33. $u^2 - 2u = u(u - 2)$

37. $14r^4 s^2 - 21rst = 7rs(2r^3 s - 3t)$

41. Can be factored no further.

45. $x^2 + 2x - 3 = (x+3)(x-1)$

49. $x^2 + 3x - 28 = (x+7)(x-4)$

53. $x^2 + 2xy + 3xz + 6yz = x(x+2y) + 3z(x+2y) = (x+2y)(x+3z).$

57. We notice that the only factors of 24 whose sum is -10 are -6 and -4. Therefore,

$$
B^2 - 10B + 24 = (B - 6)(B - 4).
$$

61. This example is factored as the difference of perfect squares. Thus,

$$
\begin{aligned}
(t+3)^2 - 16 &= ((t+3) - 4)((t+3) + 4) \\
&= (t-1)(t+7).
\end{aligned}
$$

Alternatively, we could arrive at the same answer by multiplying the expression out and then factoring it.

65.

$$
\begin{aligned}
c^2 d^2 - 25c^2 - 9d^2 + 225 &= c^2(d^2 - 25) - 9(d^2 - 25) \\
&= (d^2 - 25)(c^2 - 9) \\
&= (d+5)(d-5)(c+3)(c-3).
\end{aligned}
$$

69. The common factor is xe^{-3x}. Therefore,

$$
x^2 e^{-3x} + 2xe^{-3x} = xe^{-3x}(x + 2).
$$

73. $x^2 - 6x + 9 - 4z^2 = (x-3)^2 - (2z)^2 = (x - 3 + 2z)(x - 3 - 2z).$

77.

$$
\begin{aligned}
x^2 + 7x + 6 &= 0 \\
(x+6)(x+1) &= 0 \\
x + 6 = 0 \quad &\text{or} \quad x + 1 = 0 \\
x = -6 \quad &\text{or} \quad x = -1
\end{aligned}
$$

81.

$$
\begin{aligned}
\frac{2}{x} + \frac{3}{2x} &= 8 \\
\frac{4+3}{2x} &= 8 \\
16x &= 7 \\
x &= \frac{7}{16}
\end{aligned}
$$

85.
$$\sqrt{2x-1}+3=9$$
$$\sqrt{2x-1}=6$$
$$2x-1=36$$
$$2x=37$$
$$x=\frac{37}{2}$$

89. Rewrite the equation $g^3-4g=3g^2-12$ with a zero on the right side and factor completely.

$$g^3-3g^2-4g+12=0$$
$$g^2(g-3)-4(g-3)=0$$
$$(g-3)(g^2-4)=0$$
$$(g-3)(g+2)(g-2)=0.$$

So, $g-3=0$, $g+2=0$, or $g-2=0$. Thus, $g=3,-2$, or 2.

93. Do not divide both sides by t, because you would lose the solution $t=0$ in that case. Instead, set one side $=0$ and factor.

$$\frac{1}{64}t^3=t$$
$$\frac{1}{64}t^3-t=0$$
$$t(\frac{1}{64}t^2-1)=0$$
$$t=0 \text{ or } \frac{1}{64}t^2-1=0$$

The second equation still needs to be solved for t:

$$\frac{1}{64}t^2-1=0$$
$$\frac{1}{64}t^2=1$$
$$t^2=64$$
$$t=\pm8.$$

So the final answer is $t=0$ or $t=8$ or $t=-8$.

97. Rewrite the equation $n^5+80=5n^4+16n$ with a zero on the right side and factor completely.

$$n^5-5n^4-16n+80=0$$
$$n^4(n-5)-16(n-5)=0$$
$$(n-5)(n^4-16)=0$$
$$(n-5)(n^2-4)(n^2+4)=0$$
$$(n-5)(n+2)(n-2)(n^2+4)=0.$$

So, $n-5=0$, $n+2=0$, $n-2=0$, or $n^2+4=0$. Note that $n^2+4=0$ has no real solutions, so, $n=5,-2$, or 2.

101. First we combine like terms in the numerator.

$$\frac{x^2+1-2x^2}{(x^2+1)^2}=0$$
$$\frac{-x^2+1}{(x^2+1)^2}=0$$
$$-x^2+1=0$$

$$-x^2 = -1$$
$$x^2 = 1$$
$$x = \pm 1$$

105. We can solve this equation by cubing both sides of this equation.

$$\frac{1}{\sqrt[3]{x}} = -2$$
$$\left(\frac{1}{\sqrt[3]{x}}\right)^3 = (-2)^3$$
$$\frac{1}{x} = -8$$
$$x = -\frac{1}{8}$$

109. We begin by squaring both sides of the equation in order to eliminate the radical.

$$T = 2\pi\sqrt{\frac{l}{g}}$$
$$T^2 = 4\pi^2\left(\frac{l}{g}\right)$$
$$\frac{gT^2}{4\pi^2} = l$$

113. We substitute -3 for y in the first equation.

$$y = 2x - x^2$$
$$-3 = 2x - x^2$$
$$x^2 - 2x - 3 = 0$$
$$(x - 3)(x + 1) = 0$$
$$x = 3 \quad \text{and} \quad y = 2(3) - 3^2 = -3 \quad \text{or}$$
$$x = -1 \quad \text{and} \quad y = 2(-1) - (-1)^2 = -3$$

117. These equations cannot be solved exactly. A calculator gives the solutions as

$$x = 2.081, \quad y = 8.013 \quad \text{and} \quad x = 4.504, \quad y = 90.348.$$

CHAPTER THREE

Solutions for Section 3.1

Exercises

1. The population is growing at a rate of 1.9% per year. So, at the end of each year, the population is $100\% + 1.9\% = 101.9\%$ of what it had been the previous year. The growth factor is 1.019. If P is the population of this country, in millions, and t is the number of years since 1999, then, after one year,

$$P = 70(1.019).$$

$$\text{After two years,} \quad P = 70(1.019)(1.019) = 70(1.019)^2$$

$$\text{After three years,} \quad P = 70(1.019)(1.019)(1.019) = 70(1.019)^3$$

$$\text{After } t \text{ years,} \quad P = 70 \underbrace{(1.019)(1.019)\ldots(1.019)}_{t \text{ times}} = 70(1.019)^t$$

5. The decennial growth factor (growth factor per 10 years) is 1+ the growth per decade: 1.28.

9. For a 10% increase, we multiply by 1.10 to obtain $500 \cdot 1.10 = 550$.

13. For a 42% increase, we multiply by 1.42 to obtain $500 \cdot 1.42 = 710$. For a 42% decrease, we multiply by $1 - 0.42 = 0.58$ to obtain $710 \cdot 0.58 = 411.8$.

Problems

17. **(a)** Since N is growing by 8% per year, we know that N is an exponential function of t with growth factor $1 + 0.08 = 1.08$. Since $N = 9.8$ when $t = 0$, we have

$$N = 9.8(1.08)^t.$$

(b) In the year 2010, we have $t = 10$ and

$$N = 9.8(1.08)^{10} = 21.16 \text{ million passengers.}$$

In the year 1990, we have $t = -10$ and

$$N = 9.8(1.08)^{-10} = 4.54 \text{ million passengers.}$$

21. To find a formula for $f(n)$, we start with the number of people infected in 1990, namely P_0. In 1991, only 80% as many people, or $0.8P_0$, were infected. In 1992, again only 80% as many people were infected, which means that 80% of $0.8P_0$ people, or $0.8(0.8P_0)$ people, were infected. Continuing this line of reasoning, we can write

$$f(0) = P_0$$
$$f(1) = \underbrace{(0.80)}_{\substack{\text{one 20\%} \\ \text{reduction}}} P_0 = (0.8)^1 P_0$$

$$f(2) = \underbrace{(0.80)(0.80)}_{\substack{\text{two 20\%} \\ \text{reductions}}} P_0 = (0.8)^2 P_0$$

$$f(3) = \underbrace{(0.80)(0.80)(0.80)}_{\text{three 20\% reductions}} P_0 = (0.8)^3 P_0,$$

and so on, so that n years after 1990 we have

$$f(n) = \underbrace{(0.80)(0.80) \cdots (0.80)}_{n \text{ 20\% reductions}} P_0 = (0.8)^n P_0.$$

We see from its formula that $f(n)$ is an exponential function, because it is of the form $f(n) = ab^n$, with $a = P_0$ and $b = 0.8$. The graph of $y = f(n) = P_0(0.8)^n$, for $n \geq 0$, is given in Figure 3.1. Beginning at the P-axis, the curve decreases sharply at first towards the horizontal axis, but then levels off so that its descent is less rapid.

Figure 3.1 shows that the prevalence of the virus in the population drops quickly at first, and that it eventually levels off and approaches zero. The curve has this shape because in the early years of the vaccination program, there was a relatively large number of infected people. In later years, due to the success of the vaccine, the infection became increasingly rare. Thus, in the early years, a 20% drop in the infected population represented a larger number of people than a 20% drop in later years.

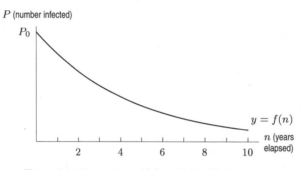

Figure 3.1: The graph of $f(n) = P_0(0.8)^n$ for $n \geq 0$

25. (a) The initial dose equals the amount of drug in the body when $t = 0$. We have $A(0) = 25(0.85)^0 = 25(1) = 25$ mg.
   (b) According to the formula,

$$A(0) = 25(0.85)^0 = 25$$
$$A(1) = 25(0.85)^1 = 25(0.85)$$
$$A(2) = 25(0.85)^2 = 25(0.85)(0.85)$$

   After each hour, the amount of the drug in the body is the amount at the end of the previous hour multiplied by 0.85. In other words, the amount remaining is 85% of what it had been an hour ago. So, 15% of the drug has left in that time.
   (c) After 10 hours, $t = 10$. $A(10) = 4.922$ mg.
   (d) Using trial and error, substitute integral values of t into $A(t) = 25(0.85)^t$ to determine the smallest value of t for which $A(t) < 1$. We find that $t = 20$ is the best choice. So, after 20 hours there will be less than one milligram in the body.

29. (a) The monthly payment on $1000 each month at 8% for a loan period of 15 years is $9.56. For $60,000, the payment would be $9.56 × 60 = $573.60 per month.
   (b) The monthly payment on $1000 each month at 8% for a loan period of 30 years is $7.34. For $60,000, the payment would be $7.34 × 60 = $440.40 per month.
   (c) The monthly payment on $1000 each month at 10% for a loan period of 15 years is $10.75. For $60,000, the payment would be $10.75 × 60 = $645.00 per month.

(d) As calculated in part (a), the monthly payment on a $60,000 loan at 8% for 15 years would be $573.60 per month. In part (c) we showed that the the monthly payment on a $60,000 loan at 10% for 15 years would be $645.00 per month. So taking the loan out at 8% rather that 10% would save the difference:

$$\text{Amount saved} = \$645.00 - \$573.60 = \$71.40 \text{ per month}$$

Since there are $15 \times 12 = 180$ months in 15 years,

$$\text{Total amount saved} = \$71.40 \text{ per month} \times 180 \text{ months} = \$12,852.$$

(e) In part (a) we found the monthly payment on an 8% mortgage of $60,000 for 15 years to be $573.60. The total amount paid over 15 years is then

$$\$573.60 \text{ per month} \times 180 \text{ months} = \$103,248.$$

In part (b) we found the monthly payment on an 8% mortgage of $60,000 for 30 years to be $440.40. The total amount paid over 30 years is then

$$440.40 \text{ per month} \times 360 \text{ months} = \$158,544.$$

The amount saved by taking the mortgage over a shorter period of time is the difference:

$$\$158,544 - \$103,248 = \$55,296.$$

33. According to the formula, we see that the growth factor is 1.047, so each year the population is 104.7% of what it was the year before, so it is growing at 4.7% per year.

If m is the monthly growth factor then, at the end of the year, it is m^{12} times the size that it was at the beginning of the year. But we know that it is 104.7% of its size at the end of the year, so

$$m^{12} = 1.047$$
$$m = 1.047^{1/12} \approx 1.00383.$$

From this we see that the monthly growth rate is approximately 0.383%.

At the end of a decade, 10 years, we know that the population is $(1.047)^{10}$ or 1.58295 times what it had been. This tells us that the growth rate per decade is 58.295%.

Solutions for Section 3.2

Exercises

1. If a function is linear and the x-values are equally spaced, you get from one y-value to the next by adding (or subtracting) the same amount each time. On the other hand, if the function is exponential and the x-values are evenly spaced, you get from one y-value to the next by multiplying by the same factor each time.

5. The formula $P_A = 200 + 1.3t$ for City A shows that its population is growing linearly. In year $t = 0$, the city has 200,000 people and the population grows by 1.3 thousand people, or 1,300 people, each year.

The formulas for cities B, C, and D show that these populations are changing exponentially. Since $P_B = 270(1.021)^t$, City B starts with 270,000 people and grows at an annual rate of 2.1%. Similarly, City C starts with 150,000 people and grows at 4.5% annually.

Since $P_D = 600(0.978)^t$, City D starts with 600,000 people, but its population decreases at a rate of 2.2% per year. We find the annual percent rate by taking $b = 0.978 = 1 + r$, which gives $r = -0.022 = -2.2\%$. So City D starts out with more people than the other three but is shrinking.

Figure 3.2 gives the graphs of the three exponential populations. Notice that the P-intercepts of the graphs correspond to the initial populations (when $t = 0$) of the towns. Although the graph of P_C starts below the graph of P_B, it eventually catches up and rises above the graph of P_B, because City C is growing faster than City B.

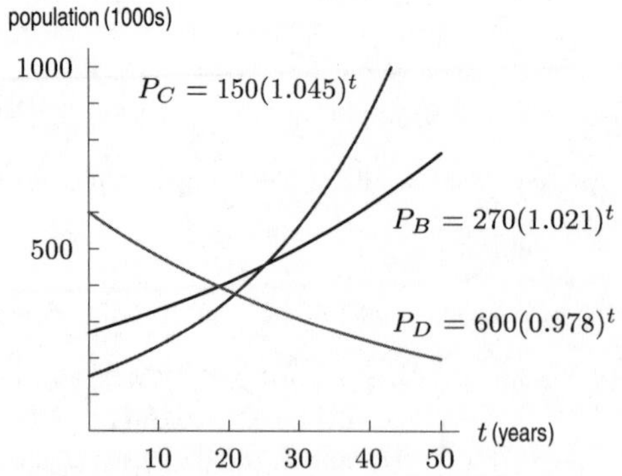

population (1000s)
1000
500
$P_C = 150(1.045)^t$
$P_B = 270(1.021)^t$
$P_D = 600(0.978)^t$
t (years)
10
20
30
40
50

Figure 3.2: The graphs of the three exponentially changing populations

9. **(a)** If a function is linear, then the differences in successive function values will be constant. If a function is exponential, the ratios of successive function values will remain constant. Now

$$i(1) - i(0) = 14 - 18 = -4$$

and

$$i(2) - i(1) = 10 - 14 = -4.$$

Checking the rest of the data, we see that the differences remain constant, so $i(x)$ is linear.

(b) We know that $i(x)$ is linear, so it must be of the form

$$i(x) = b + mx,$$

where m is the slope and b is the y-intercept. Since at $x = 0$, $i(0) = 18$, we know that the y-intercept is 18, so $b = 18$. Also, we know that at $x = 1$, $i(1) = 14$, we have

$$\begin{aligned} i(1) &= b + m \cdot 1 \\ 14 &= 18 + m \\ m &= -4. \end{aligned}$$

Thus, $i(x) = 18 - 4x$. The graph of $i(x)$ is shown in Figure 3.3.

y
18
10
2
x
1
2
3
4

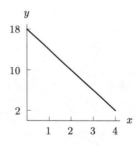

Figure 3.3

Problems

13. Since $g(x) = ab^x$, we can say that $g(\frac{1}{2}) = ab^{1/2}$ and $g(\frac{1}{4}) = ab^{1/4}$. Since we know that $g(\frac{1}{2}) = 4$ and $g(\frac{1}{4}) = 2\sqrt{2}$, we can conclude that

$$ab^{1/2} = 4 = 2^2$$

and

$$ab^{1/4} = 2\sqrt{2} = 2 \cdot 2^{1/2} = 2^{3/2}.$$

Forming ratios, we have

$$\frac{ab^{1/2}}{ab^{1/4}} = \frac{2^2}{2^{3/2}}$$
$$b^{1/4} = 2^{1/2}$$
$$(b^{1/4})^4 = (2^{1/2})^4$$
$$b = 2^2 = 4.$$

Now we know that $g(x) = a(4)^x$, so $g(\frac{1}{2}) = a(4)^{1/2} = 2a$. Since we also know that $g(\frac{1}{2}) = 4$, we can say

$$2a = 4$$
$$a = 2.$$

Therefore $g(x) = 2(4)^x$.

17. Since the function is exponential, we know that $y = ab^x$. Since $(0, 1.2)$ is on the graph, we know $1.2 = ab^0$, and that $a = 1.2$. To find b, we use point $(2, 4.8)$ which gives

$$4.8 = 1.2(b)^2$$
$$4 = b^2$$
$$b = 2, \text{ since } b > 0.$$

Thus, $y = 1.2(2)^x$ is a possible formula for this function.

21. To use the ratio method we must have the y-values given at equally spaced x-values, which they are not. However, some of them are spaced 1 apart, namely, 1 and 2; 4 and 5; and 8 and 9. Thus, we can use these values, and consider

$$\frac{f(2)}{f(1)}, \frac{f(5)}{f(4)}, \text{ and } \frac{f(9)}{f(8)}.$$

We find

$$\frac{f(2)}{f(1)} = \frac{f(5)}{f(4)} = \frac{f(9)}{f(8)} = \frac{1}{4}.$$

With $f(x) = ab^x$ we also have

$$\frac{f(2)}{f(1)} = \frac{f(5)}{f(4)} = \frac{f(9)}{f(8)} = b,$$

so $b = \frac{1}{4}$. Using $f(1) = 4096$ we find $4096 = ab = a\left(\frac{1}{4}\right)$, so $a = 16,384$. Thus, $f(x) = 16,384\left(\frac{1}{4}\right)^x$.

25. This table could represent an exponential function, since for every Δt of 1, the value of $g(t)$ halves. This means that b in the form $g(t) = ab^t$ must be $\frac{1}{2} = 0.5$. We can solve for a by substituting in $(1, 512)$ (or any other point):

$$512 = a \cdot 0.5^1$$
$$512 = a \cdot 0.5$$
$$1024 = a.$$

Thus, a possible formula to describe the data in the table is $g(t) = 1024 \cdot 0.5^t$.

29. (a) If P is linear, then $P(t) = b + mt$ and

$$m = \frac{\Delta P}{\Delta t} = \frac{P(13) - P(7)}{13 - 7} = \frac{3.75 - 3.21}{13 - 7} = \frac{0.54}{6} = 0.09.$$

So $P(t) = b + 0.09t$ and $P(7) = b + 0.09(7)$. We can use this and the fact that $P(7) = 3.21$ to say that

$$3.21 = b + 0.09(7)$$
$$3.21 = b + 0.63$$
$$2.58 = b.$$

So $P(t) = 2.58 + 0.09t$. The slope is 0.09 million people per year. This tells us that, if its growth is linear, the country grows by $0.09(1,000,000) = 90,000$ people every year.

(b) If P is exponential, $P(t) = ab^t$. So

$$P(7) = ab^7 = 3.21$$

and

$$P(13) = ab^{13} = 3.75.$$

We can say that

$$\frac{P(13)}{P(7)} = \frac{ab^{13}}{ab^7} = \frac{3.75}{3.21}$$
$$b^6 = \frac{3.75}{3.21}$$
$$(b^6)^{1/6} = \left(\frac{3.75}{3.21}\right)^{1/6}$$
$$b = 1.026.$$

Thus, $P(t) = a(1.026)^t$. To find a, note that

$$P(7) = a(1.026)^7 = 3.21$$
$$a = \frac{3.21}{(1.026)^7} = 2.68.$$

We have $P(t) = 2.68(1.026)^t$. Since $b = 1.026$ is the growth factor, the country's population grows by about 2.6% per year, assuming exponential growth.

33. (a) We want $N = f(t)$ so we have

$$\text{Slope} = \frac{\Delta N}{\Delta t} = \frac{130 - 84}{2001 - 1990} = \frac{46}{11} = 4.182.$$

Since $N = 84$ when $t = 0$, the vertical intercept is 84 and the linear formula is

$$N = 84 + 4.182t.$$

The slope is 4.182. The number of asthma sufferers has increased, on average, by 4.182 million people per year during this period.

(b) Since $N = 84$ when $t = 0$, we have the exponential function $N = 84b^t$ for some base b. Since $N = 130$ when $t = 11$, we have

$$N = 84b^t$$
$$130 = 84b^{11}$$
$$b^{11} = \frac{130}{84} = 1.5476$$
$$b = (1.5476)^{1/11} = 1.0405.$$

The exponential formula is

$$N = 84(1.0405)^t.$$

The growth factor is 1.0405. The number of asthma sufferers has increased, on average, by 4.05% per year during this period.

(c) In the year 2010, we have $t = 20$. Using the linear formula, the predicted number in 2010 is

$$N = 84 + 4.182(20) = 167.640 \text{ million asthma sufferers.}$$

Using the exponential formula, the predicted number in 2010 is

$$N = 84(1.0405)^{20} = 185.832 \text{ million asthma sufferers.}$$

37. (a) (v) In $k(x) = A(2)^{-x} = A(1/2)^x$, A would be the initial value (0.35). The $(2)^{-x} = 1/2^x$ term tells us that the function is decreasing by half each year.

(b) (iii) In $h(x) = B(0.7)^x$, B is the initial charge. The $(0.7)^x$ term tells us that at the end of each second, the amount of charge is 70% of what it had been the previous second. Therefore, it has decreased by 30%.

(c) (iv) In $j(x) = B(0.3)^x$, B is the initial level of pollutants. The $(0.3)^x = (0.30)^x$ term tells us that 30% of the pollutants remain after each filter.

(d) (ii) In $g(x) = P_0(1 + r)^x$, $P_0 = 3000$ represents the initial population. The $(1 + r)^x$ term represents the growth factor, with $r = 0.10$, a 10% increase, and $0 \le x \le 5$ since there are 5 decades between 1920 and 1970.

(e) (i) In $f(x) = P_0 + rx$, $P_0 = 3000$ is the initial population, $r = 250$ is the number by which the town grew every year and $0 \le x \le 50$.

Solutions for Section 3.3

Exercises

1. Since $y = a$ when $t = 0$ in $y = ab^t$, a is the y-intercept. Thus, the function with the greatest y-intercept, D, has the largest a.

5. Graphing $y = 46(1.1)^x$ and tracing along the graph on a calculator gives us an answer of $x = 7.158$. See Figure 3.4.

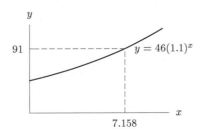

Figure 3.4

9. As t approaches $-\infty$, the value of ab^t approaches zero for any a.

Problems

13. Let $f(x) = (1.1)^x$, $g(x) = (1.2)^x$, and $h(x) = (1.25)^x$. We note that for $x = 0$,

$$f(x) = g(x) = h(x) = 1;$$

so all three graphs have the same y-intercept. On the other hand, for $x = 1$,

$$f(1) = 1.1, \quad g(1) = 1.2, \quad \text{and} \quad h(1) = 1.25,$$

so $0 < f(1) < g(1) < h(1)$. For $x = 2$,

$$f(2) = 1.21, \quad g(2) = 1.44, \quad \text{and} \quad h(2) = 1.5625,$$

so $0 < f(2) < g(2) < h(2)$. In general, for $x > 0$,

$$0 < f(x) < g(x) < h(x).$$

This suggests that the graph of $f(x)$ lies below the graph of $g(x)$, which in turn lies below the graph of $h(x)$, and that all lie above the x-axis. Alternately, you can consider 1.1, 1.2, and 1.25 as growth factors to conclude $h(x) = (1.25)^x$ is the top function, and $g(x) = (1.2)^x$ is in the middle, $f(x)$ is at the bottom.

17. (a) Since the number of cases is reduced by 10% each year, there are 90% as many cases in one year as in the previous one. So, after one year there are 90% of 10000 or 10000(0.90) cases, while after two years, there are $10000(0.90)(0.90) = 10000(0.90)^2$ cases. In general, the number of cases after t years is $y = (10000)(0.9)^t$.

(b) Setting $t = 5$, we obtain the number of cases 5 years from now

$$y = (10000) \cdot (0.9)^5 = 5904.9 \approx 5905 \text{ cases.}$$

(c) Plotting $y = (10000) \cdot (0.9)^t$ and approximating the value of t for which $y = 1000$, we obtain $t \approx 21.854$ years.

21. Increasing: $b > 1, a > 0$ or $0 < b < 1, a < 0$;
Decreasing: $0 < b < 1, a > 0$ or $b > 1, a < 0$;
The function is concave up for $a > 0, 0 < b < 1$ or $b > 1$.

25. A possible graph is shown in Figure 3.5.

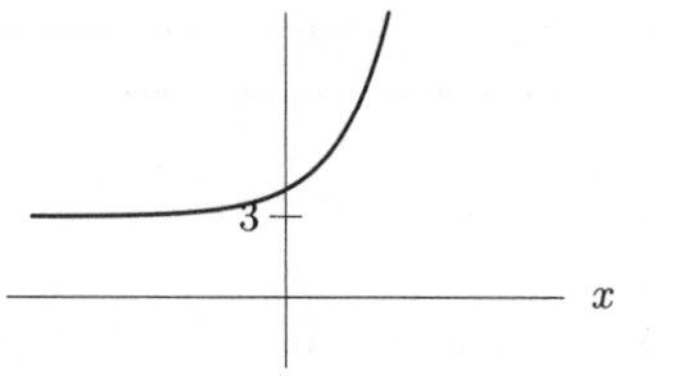

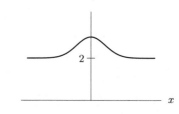

Figure 3.5

Figure 3.6

29. We see in Figure 3.6 that this function has a horizontal asymptote of $y = 2$.

33. (a) One algorithm used by a calculator or computer gives the exponential regression function as

$$S = 92(1.20)^t$$

Other algorithms may give different formulas.

(b) See Figure 3.7. The exponential function appears to fit the points reasonably well.

(c) Since the base of this exponential function is 1.20, sales have been increasing at a rate of about 20% per year during this period.

(d) Using $t = 20$, we have $S = 92(1.20)^{20} = 3527$ million sales.

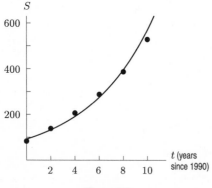

Figure 3.7

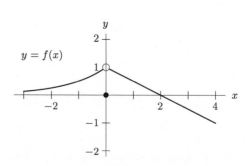

Figure 3.8

37. (a) See Figure 3.8

(b) The range of this function is all real numbers less than one — i.e. $f(x) < 1$.

(c) The y-intercept occurs at $(0,0)$. This point is also an x-intercept. To solve for other x-intercepts we must attempt to solve $f(x) = 0$ for each of the two remaining parts of f. In the first case, we know that the function $f(x) = 2^x$ has no x-intercepts, as there is no value of x for which 2^x is equal to zero. In the last case, for $x > 0$, we set $f(x) = 0$ and solve for x:

$$0 = 1 - \frac{1}{2}x$$

$$\frac{1}{2}x = 1$$

$$x = 2.$$

Hence $x = 2$ is another x-intercept of f.

(d) As x gets large, the function is defined by $f(x) = 1 - 1/2x$. To determine what happens to f as $x \to +\infty$, find values of f for very large values of x. For example,

$$f(100) = 1 - \frac{1}{2}(100) = -49, \quad f(10000) = 1 - \frac{1}{2}(10000) = -4999$$

$$\text{and} \quad f(1{,}000{,}000) = 1 - \frac{1}{2}(1{,}000{,}000) = -499{,}999.$$

As x becomes larger, $f(x)$ becomes more and more negative. A way to write this is:

$$\text{As } x \to +\infty, \ f(x) \to -\infty.$$

As x gets very negative, the function is defined by $f(x) = 2^x$.

Choosing very negative values of x, we get $f(-100) = 2^{-100} = 1/2^{100}$, and $f(-1000) = 2^{-1000} = 1/2^{1000}$. As x becomes more negative the function values get closer to zero. We write

$$\text{As } x \to -\infty, \ f(x) \to 0.$$

(e) Increasing for $x < 0$, decreasing for $x > 0$.

Solutions for Section 3.4

Exercises

1. We know that $e \approx 2.71828$, so $2 < e < 3$. Since e lies between 2 and 3, the graph of $y = e^x$ lies between the graphs of $y = 2^x$ and $y = 3^x$. Since 3^x increases faster than 2^x, the correct matching is shown in Figure 3.9.

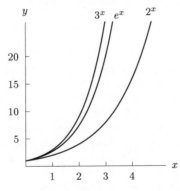

Figure 3.9

5. With continuous compounding, the interest earns interest during the year, so the balance grows faster with continuous compounding than with annual compounding. Curve A corresponds to continuous compounding and curve B corresponds to annual compounding. The initial amount in both cases is the vertical intercept, $500.

9. (a) If the interest is compounded annually, there will be $500 \cdot 1.05 = \$525$ after one year.
   (b) If the interest is compounded weekly, there will be $500 \cdot (1 + 0.05/52)^{52} = \$525.62$ after one year.
   (c) If the interest is compounded every minute, there will be $500 \cdot (1 + 0.05/525{,}600)^{525{,}600} = \$525.64$ after one year.
   (d) If the interest is compounded continuously, there will be $500 \cdot e^{0.05} = \$525.64$ after one year.

Problems

13. Using the formula $y = ab^x$, each of the functions has the same value for b, but different values for a and thus different y-intercepts.
   When $x = 0$, the y-intercept for $y = e^x$ is 1 since $e^0 = 1$.
   When $x = 0$, the y-intercept for $y = 2e^x$ is 2 since $e^0 = 1$ and $2(1) = 2$.
   When $x = 0$, the y-intercept for $y = 3e^x$ is 3 since $e^0 = 1$ and $3(1) = 3$.
   Therefore, $y = e^x$ is the bottom graph, above it is $y = 2e^x$ and the top graph is $y = 3e^x$.

17. We want to know when $V = 1074$. Tracing along the graph of $V = 537e^{0.015t}$ gives $t \approx 46.210$ years.

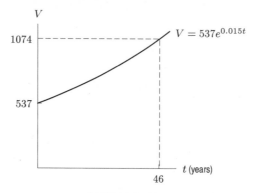

Figure 3.10

21. (a) (i) $B = B_0 \left(1 + \dfrac{.06}{4}\right)^4 \approx B_0(1.0614)$, so the APR is approximately 6.14%.

   (ii) $B = B_0 \left(1 + \dfrac{.06}{12}\right)^{12} \approx B_0(1.0617)$, so the APR is approximately 6.17%.

   (iii) $B = B_0 \left(1 + \dfrac{.06}{52}\right)^{52} \approx B_0(1.0618)$, so the APR is approximately 6.18%.

   (iv) $B = B_0 \left(1 + \dfrac{.06}{365}\right)^{365} \approx B_0(1.0618)$, so the APR is approximately 6.18%.

   (b) $e^{0.06} \approx 1.0618$. No matter how often we compound interest, we'll never get more than $\approx 6.18\%$ APR.

25. (a) The nominal rate is the stated annual interest without compounding, thus 6%.
      The effective annual rate for an account paying 1% compounded annually is 6%.
   (b) The nominal rate is the stated annual interest without compounding, thus 6%.
      With quarterly compounding, there are four interest payments per year, each of which is $6/4 = 1.5\%$. Over the course of the year, this occurs four times, giving an effective annual rate of $1.015^4 = 1.06136$, which is 6.136%.
   (c) The nominal rate is the stated annual interest without compounding, thus 6%.
      With daily compounding, there are 365 interest payments per year, each of which is $(6/365)\%$. Over the course of the year, this occurs 365 times, giving an effective annual rate of $(1 + 0.06/365)^{365} = 1.06183$, which is 6.183%.
   (d) The nominal rate is the stated annual interest without compounding, thus 6%.
      The effective annual rate for an account paying 6% compounded continuously is $e^{0.06} = 1.06184$, which is 6.184%.

29. To see which investment is best after 1 year, we compute the effective annual yield:

For Bank A, $P = P_0(1 + \frac{0.07}{365})^{365(1)} \approx 1.0725P_0$

For Bank B, $P = P_0(1 + \frac{0.071}{12})^{12(1)} \approx 1.0734P_0$

For Bank C, $P = P_0(e^{0.0705(1)}) \approx 1.0730P_0$

Therefore, the best investment is with Bank B, followed by Bank C and then Bank A.

33. (a) The effective annual rate is the rate at which the account is actually increasing in one year. According to the formula, $M = M_0(1.07763)^t$, at the end of one year you have $M = 1.07763M_0$, or 1.07763 times what you had the previous year. The account is 107.763% larger than it had been previously; that is, it increased by 7.763%. Thus the effective rate being paid on this account each year is about 7.763%.

(b) Since the money is being compounded each month, one way to find the nominal annual rate is to determine the rate being paid each month. In t years there are $12t$ months, and so, if b is the monthly growth factor, our formula becomes

$$M = M_0 b^{12t} = M_0 (b^{12})^t.$$

Thus, equating the two expressions for M, we see that

$$M_0 (b^{12})^t = M_0 (1.07763)^t.$$

Dividing both sides by M_0 yields

$$(b^{12})^t = (1.07763)^t.$$

Taking the tth root of both sides, we have

$$b^{12} = 1.07763$$

which means that

$$b = (1.07763)^{1/12} \approx 1.00625.$$

Thus, this account earns 0.625% interest every month, which amounts to a nominal interest rate of about $12(0.625\%) = 7.5\%$.

Solutions for Chapter 3 Review

Exercises

1. If the population is growing or shrinking at a constant rate of m people per year, the formula is linear. Since the vertical intercept is 3000, we have $P = 3000 + mt$.

If the population is growing or shrinking at a constant percent rate of r percent per year, the formula is exponential in the form $P = a(1 + r)^t$. Since the vertical intercept is 3000, we have $P = 3000(1 + r)^t$.

If the population is growing or shrinking at a constant continuous percent rate of k percent per year, the formula is exponential in the form $P = ae^{kt}$. Since the vertical intercept is 3000, we have $P = 3000e^{kt}$.

We have:

(a) $P = 3000 + 200t$.

(b) $P = 3000(1.06)^t$.

(c) $P = 3000e^{0.06t}$.

(d) $P = 3000 - 50t$.

(e) $P = 3000(0.96)^t$.

(f) $P = 3000e^{-0.04t}$.

5. To match formula and graph, we keep in mind the effect on the graph of the parameters a and b in $y = ab^t$.

If $a > 0$ and $b > 1$, then the function is positive and increasing.

If $a > 0$ and $0 < b < 1$, then the function is positive and decreasing.

If $a < 0$ and $b > 1$, then the function is negative and decreasing.

If $a < 0$ and $0 < b < 1$, then the function is negative and increasing.

(a) $y = 0.8^t$. So $a = 1$ and $b = 0.8$. Since $a > 0$ and $0 < b < 1$, we want a graph that is positive and decreasing. The graph in (ii) satisfies the conditions.

(b) $y = 5(3)^t$. So $a = 5$ and $b = 3$. The graph in (i) is both positive and increasing.

(c) $y = -6(1.03)^t$. So $a = -6$ and $b = 1.03$. Here, $a < 0$ and $b > 1$, so we need a graph which is negative and decreasing. The graph in (iv) satisfies these conditions.

(d) $y = 15(3)^{-t}$. Since $(3)^{-t} = (3)^{-1 \cdot t} = (3^{-1})^t = (\frac{1}{3})^t$, this formula can also be written $y = 15(\frac{1}{3})^t$. $a = 15$ and $b = \frac{1}{3}$. A graph that is both positive and decreasing is the one in (ii).

(e) $y = -4(0.98)^t$. So $a = -4$ and $b = 0.98$. Since $a < 0$ and $0 < b < 1$, we want a graph which is both negative and increasing. The graph in (iii) satisfies these conditions.

(f) $y = 82(0.8)^{-t}$. Since $(0.8)^{-t} = (\frac{8}{10})^{-t} = (\frac{8}{10})^{-1 \cdot t} = ((\frac{8}{10})^{-1})^t = (\frac{10}{8})^t = (1.25)^t$ this formula can also be written as $y = 82(1.25)^t$. So $a = 82$ and $b = 1.25$. A graph which is both positive and increasing is the one in (i).

9. Since the function is exponential, we know $y = ab^x$. We also know that $(0, 1/2)$ and $(3, 1/54)$ are on the graph of this function, so $1/2 = ab^0$ and $1/54 = ab^3$. The first equation implies that $a = 1/2$. Substituting this value in the second equation gives $1/54 = (1/2)b^3$ or $b^3 = 1/27$, or $b = 1/3$. Thus, $y = \dfrac{1}{2}\left(\dfrac{1}{3}\right)^x$.

13. As the x values go up by 1, the corresponding $g(x)$ values go down by 1.3 each time, so the function might be linear with slope -1.3. Since $g(x) = 12.8$ when $x = 0$, we have

$$g(x) = 12.8 - 1.3x.$$

17. Since f is exponential, $f(x) = ab^x$. We know that

$$f(2) = ab^2 = \frac{2}{9}$$

and

$$f(-3) = ab^{-3} = 54,$$

so

$$\frac{2/9}{54} = \frac{ab^2}{ab^{-3}} = b^5.$$

$$b^5 = \frac{1}{243}$$

$$b = \left(\frac{1}{243}\right)^{1/5} = \frac{1}{3}.$$

Thus, $f(x) = a\left(\dfrac{1}{3}\right)^x$. Since $f(2) = \dfrac{2}{9}$ and $f(2) = a(\frac{1}{3})^2$, we have

$$a\left(\frac{1}{3}\right)^2 = \frac{2}{9}$$

$$\frac{a}{9} = \frac{2}{9}$$

$$a = 2.$$

Thus, $f(x) = 2\left(\frac{1}{3}\right)^x$.

21. A graph of $y = 3 - e^{-x}$ is shown in Figure 3.11. We see that there is a horizontal asymptote at $y = 3$. We can also see this by noting that as $x \to \infty$, we have $e^{-x} \to 0$ and so $3 - e^{-x} \to 3$.

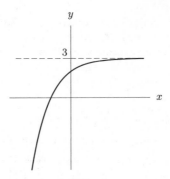

Figure 3.11

Problems

25. If P represents population and t is the number of years since 2001, then in 2001, where $t = 0$ and $P = 284.8$ million. If the population increases by 0.9% per year, then, each year, it is 100.9% of what it had been the year before. So we know that $P = 284.8(1.009)^t$. We want to know t when $P = 300$ million, so we solve $300 = 284.8(1.009)^t$. Using a graph or trial and error calculation, we project that for $t \approx 6$ years after 2001, or approximately in the year 2007, the population will have risen to 300 million.

29. $f(x)$ is not exponential, since $f(0) = 0$, and exponential functions do not intersect the x-axis. Next, we calculate rates of change:

$$\frac{f(1) - f(0)}{1 - 0} = \frac{1 - 0}{1} = 1$$

and

$$\frac{f(-1) - f(-2)}{-1 - (-2)} = \frac{1 - 4}{-1 + 2} = -3.$$

Since the rates of change are not constant, we know that $f(x)$ is not linear either.

As we look at consecutive values of $g(x)$, we can see that each value is $\frac{1}{4}$ of the previous value:

$$\frac{g(1)}{g(0)} = \frac{g(0)}{g(-1)} = \frac{g(-1)}{g(-2)} = \frac{1}{4}.$$

Since the ratio between successive values of g is constant, g is exponential, and a formula for $g(x)$ is $g(x) = ab^x$. Since b is the growth factor of our function, which is $\frac{1}{4}$ in this case, we have $g(x) = a(\frac{1}{4})^x$. Since $g(0) = a(\frac{1}{4})^0 = a$ and $g(0) = 3$, we have $g(x) = 3\left(\frac{1}{4}\right)^x$.

$h(x)$ is linear, which we can confirm by noticing that the rates of change are constant:

$$\frac{h(1) - h(0)}{1 - 0} = \frac{h(0) - h(-1)}{0 - (-1)} = \frac{h(-1) - h(-2)}{(-1) - (-2)} = -\frac{4}{3}.$$

We know that $h(x) = b + mx$, and that m, the constant rate of change, is $-\frac{4}{3}$, so

$$h(x) = b - \frac{4}{3}x.$$

We also know that b is our initial value when $x = 0$. Since $h(0) = b = 4$, we have $h(x) = 4 - \frac{4}{3}x$.

33. **(a)** Accion's interest rate $= (1160 - 1000)/1000 = 0.16 = 16\%$.
   (b) Payment to loan shark $= 1000 + 22\% \cdot 1000 = \$1220$.
   (c) The one from Accion, since the interest rate is lower.

37. Let r be the percentage by which the substance decays each year. Every year we multiply the amount of radioactive substance by $1 - r$ to determine the new amount. If a is the amount of the substance on hand originally, we know that after five years, there have been five yearly decreases, by a factor of $1 - r$. Since we know that there will be 60% of a, or $0.6a$, remaining after five years (because 40% of the original amount will have decayed), we know that

$$a \cdot \underbrace{(1 - r)^5}_{\substack{\text{five annual decreases} \\ \text{by a factor of } 1 - r}} = 0.6a.$$

Dividing both sides by a, we have $(1 - r)^5 = 0.6$, which means that

$$1 - r = (0.6)^{\frac{1}{5}} \approx 0.9029$$

so

$$r \approx 0.09712 = 9.712\%.$$

Each year the substance decays by 9.712%.

41. (a) The data points are approximately as shown in Table 3.1. This results in $a \approx 15.269$ and $b \approx 1.122$, so $E(t) = 15.269(1.122)^t$.

Table 3.1

t (years)	0	1	2	3	4	5	6	7	8	9	10	11	12
$E(t)$ (thousands)	22	18	20	20	22	22	19	30	45	42	62	60	65

(b) In 1997 we have $t = 17$ so $E(17) = 15.269(1.122)^{17} \approx 108,066$.

(c) The model is probably not a good predictor of emigration in the year 2010 because Hong Kong has been transferred to Chinese rule. Thus, conditions which may affect emigration are markedly different than in the period from 1989 to 1992, for which data is given.

45. (i) Equation (b). Since the growth factor is 1.12, or 112%, the annual interest rate is 12%.

(ii) Equation (a). An account earning at least 1% monthly will have a monthly growth factor of at least 1.01, which means that the annual (12-month) growth factor will be at least

$$(1.01)^{12} \approx 1.1268.$$

Thus, an account earning at least 1% monthly will earn *at least* 12.68%. The only account that earns this much interest is account (a).

(iii) Equation (c). An account earning 12% annually compounded semi-annually will earn 6% twice yearly. In t years, there are $2t$ half-years.

(iv) Equations (b), (c) and (d). An account that earns 3% each quarter ends up with a yearly growth factor of $(1.03)^4 \approx 1.1255$. This corresponds to an annual percentage rate of 12.55%. Accounts (b), (c) and (d) earn less than this. Check this by determining the growth factor in each case.

(v) Equations (a) and (e). An account that earns 6% every 6 months will have a growth factor, after 1 year, of $(1 + 0.06)^2 = 1.1236$, which is equivalent to a 12.36% annual interest rate, compounded annually. Account (a), earning 20% each year, clearly earns more than 6% twice each year, or 12.36% annually. Account (e), which earns 3% each quarter, earns $(1.03)^2 = 1.0609$, or 6.09% every 6 months, which is greater than 6% every 2 quarters.

49. (a)

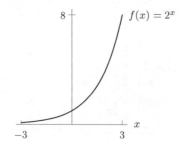

(b) The point $(0, 1)$ is on the graph. So is $(0.01, 1.00696)$. Taking $\dfrac{y_2 - y_1}{x_2 - x_1}$, we get an estimate for the slope of 0.696. We may zoom in still further to find that $(0.001, 1.000693)$ is on the graph. Using this and the point $(0, 1)$ we would get a slope of 0.693. Zooming in still further we find that the slope stabilizes at around 0.693; so, to two digits of accuracy, the slope is 0.69.

(c) Using the same method as in part (b), we find that the slope is ≈ 1.10.

(d) We might suppose that the slope of the tangent line at $x = 0$ increases as b increases. Trying a few values, we see that this is the case. Then we can find the correct b by trial and error: $b = 2.5$ has slope around 0.916, $b = 3$ has slope around 1.1, so $2.5 < b < 3$. Trying $b = 2.75$ we get a slope of 1.011, just a little too high. $b = 2.7$ gives a slope of 0.993, just a little too low. $b = 2.72$ gives a slope of 1.0006, which is as good as we can do by giving b to two decimal places. Thus $b \approx 2.72$.

In fact, the slope is exactly 1 when $b = e = 2.718\ldots$.

CHECK YOUR UNDERSTANDING

1. True. If the constant rate is r then the formula is $f(t) = a \cdot (1 + r)^t$. The function decreases when $0 < 1 + r < 1$ and increases when $1 + r > 1$.

5. False. The annual growth factor would be 1.04, so $S = S_0(1.04)^t$.

9. True. The initial value means the value of Q when $t = 0$, so $Q = f(0) = a \cdot b^0 = a \cdot 1 = a$.

13. False. This is the formula of a linear function.

17. True. The irrational number $e = 2.71828\cdots$ has this as a good approximation.

21. True. The initial value is 200 and the growth factor is 1.04.

25. True. Since k is the continuous growth rate and negative, Q is decreasing.

29. True. The interest from any quarter is compounded in subsequent quarters.

Solutions to Tools for Chapter 3

1. $4^3 = 4 \cdot 4 \cdot 4 = 64$

5. $(-1)^{12} = \underbrace{(-1)(-1)\cdots(-1)}_{12 \; factors} = 1$

9. $\dfrac{10^8}{10^5} = 10^{8-5} = 10^3 = 10 \cdot 10 \cdot 10 = 1{,}000$

13. $\sqrt{4^2} = 4$

17. Since $\dfrac{1}{7^{-2}}$ is the same as 7^2, we obtain $7 \cdot 7$ or 49.

21. The order of operations tells us to square 3 first (giving 9) and then multiply by -2. Therefore $(-2)3^2 = (-2)9 = -18$.

25. $16^{1/2} = (2^4)^{1/2} = 2^2 = 4$

29. $16^{5/2} = (2^4)^{5/2} = 2^{10} = 1024$

33. Exponentiation is done first, with the result that $(-1)^3 = -1$. Therefore $(-1)^3 \sqrt{36} = (-1)\sqrt{36} = (-1)(6) = -6$.

37. $3^{-2} = \dfrac{1}{3^2} = \dfrac{1}{9}$

41. $25^{-3/2} = \dfrac{1}{(25)^{3/2}} = \dfrac{1}{(25^{1/2})^3} = \dfrac{1}{5^3} = \dfrac{1}{125}$

45. $\sqrt{y^8} = (y^8)^{1/2} = y^{8/2} = y^4$

49. $\sqrt{49w^9} = (49w^9)^{1/2} = 49^{1/2} \cdot w^{9/2} = 7w^{9/2}$

53. $\sqrt{r^4} = (r^4)^{1/2} = r^{4/2} = r^2$

57.

$$\sqrt{48u^{10}v^{12}y^5} = (48)^{1/2} \cdot (u^{10})^{1/2} \cdot (v^{12})^{1/2} \cdot (y^5)^{1/2}$$
$$= (16 \cdot 3)^{1/2} u^5 v^6 y^{5/2}$$
$$= 16^{1/2} \cdot 3^{1/2} \cdot u^5 v^6 y^{5/2}$$
$$= 4\sqrt{3}u^5 v^6 y^{5/2}$$

61. First we raise $3^{x/2}$ to the second power and multiply this result by 3. Therefore $3\left(3^{x/2}\right)^2 = 3\left(3^x\right) = 3^1\left(3^x\right) = 3^{x+1}$.

65. $\sqrt{e^{2x}} = \left(e^{2x}\right)^{\frac{1}{2}} = e^{2x \cdot \frac{1}{2}} = e^x$

69. Inside the parenthesis we write the radical as an exponent, which results in

$$\left(3x\sqrt{x^3}\right)^2 = \left(3x \cdot x^{3/2}\right)^2.$$

Then within the parenthesis we write

$$\left(3x^1 \cdot x^{3/2}\right)^2 = \left(3x^{5/2}\right)^2 = 3^2(x^{5/2})^2 = 9x^5.$$

73. $\dfrac{4A^{-3}}{(2A)^{-4}} = \dfrac{4/A^3}{1/(2A)^4} = \dfrac{4}{A^3} \cdot \dfrac{(2A)^4}{1} = \dfrac{4}{A^3} \cdot \dfrac{2^4 A^4}{1} = 64A.$

77. First we divide within the larger parentheses. Therefore,

$$\left(\frac{35(2b+1)^9}{7(2b+1)^{-1}}\right)^2 = \left(5(2b+1)^{9-(-1)}\right)^2 = \left(5(2b+1)^{10}\right)^2.$$

Then we expand to obtain

$$25(2b+1)^{20}.$$

81. $(-625)^{3/4} = (\sqrt[4]{-625})^3$. Since $\sqrt[4]{-625}$ is not a real number, $(-625)^{3/4}$ is undefined.

85. $(-64)^{3/2} = (\sqrt{-64})^3$. Since $\sqrt{-64}$ is not a real number, $(-64)^{3/2}$ is undefined.

89. We have

$$\sqrt{4x^3} = 5$$
$$2x^{3/2} = 5$$
$$x^{3/2} = 2.5$$
$$x = (2.5)^{2/3} = 1.842.$$

93. The point of intersection occurs where the curves have the same x and y values. We set the two formulas equal and solve:

$$0.8x^4 = 5x^2$$
$$\frac{x^4}{x^2} = \frac{5}{0.8}$$
$$x^2 = 6.25$$
$$x = (6.25)^{1/2} = 2.5.$$

The x coordinate of the point of intersection is 2.5. We use either formula to find the y-coordinate:

$$y = 5(2.5)^2 = 31.25,$$

or

$$y = 0.8(2.5)^4 = 31.25.$$

The coordinates of the point of intersection are $(2.5, 31.25)$.

97. False

101. False

CHAPTER FOUR

Solutions for Section 4.1

Exercises

1. The statement is equivalent to $26 = e^{3.258}$.

5. The statement is equivalent to $q = e^z$.

9. The statement is equivalent to $v = \log \alpha$.

13. We are solving for an exponent, so we use logarithms. Since the base is the number e, it makes the most sense to use the natural logarithm. Using the log rules, we have

$$e^{0.12x} = 100$$
$$\ln(e^{0.12x}) = \ln(100)$$
$$0.12x = \ln(100)$$
$$x = \frac{\ln(100)}{0.12} = 38.376.$$

17. We take the log of both sides and use the rules of logs to solve for m:

$$\log(0.00012) = \log(0.001)^{m/2}$$
$$\log(0.00012) = \frac{m}{2}\log(0.001)$$
$$\log(0.00012) = \frac{m}{2}(-3)$$
$$\frac{\log(0.00012)}{-3} = \frac{m}{2}$$
$$m = 2\frac{\log(0.00012)}{-3} = 2.614.$$

Problems

21. **(a)** Since $1 = e^0$, $\ln 1 = 0$.
 (b) Using the identity $\ln e^N = N$, we get $\ln e^0 = 0$. Or we could notice that $e^0 = 1$, so using part (a), $\ln e^0 = \ln 1 = 0$.
 (c) Using the identity $\ln e^N = N$, we get $\ln e^5 = 5$.
 (d) Recall that $\sqrt{e} = e^{1/2}$. Using the identity $\ln e^N = N$, we get $\ln \sqrt{e} = \ln e^{1/2} = \frac{1}{2}$.
 (e) Using the identity $e^{\ln N} = N$, we get $e^{\ln 2} = 2$.
 (f) Since $\frac{1}{\sqrt{e}} = e^{-1/2}$, $\ln \frac{1}{\sqrt{e}} = \ln e^{-1/2} = -\frac{1}{2}$

25. **(a)** True.
 (b) False. $\frac{\log A}{\log B}$ cannot be rewritten.
 (c) False. $\log A \log B = \log A \cdot \log B$, not $\log A + \log B$.
 (d) True.
 (e) True. $\sqrt{x} = x^{1/2}$ and $\log x^{1/2} = \frac{1}{2}\log x$.
 (f) False. $\sqrt{\log x} = (\log x)^{1/2}$.

29. **(a)** We have $\ln(nm^4) = \ln n + 4\ln m = q + 4p$.
 (b) We have $\ln(1/n) = \ln 1 - \ln n = 0 - \ln n = -q$.
 (c) We have $(\ln m)/(\ln n) = p/q$.
 (d) We have $\ln(n^3) = 3\ln n = 3q$.

33. Using the log rules, we have

$$4(1.171)^x = 7(1.088)^x$$

$$\frac{(1.171)^x}{(1.088)^x} = \frac{7}{4}$$

$$\left(\frac{1.171}{1.088}\right)^x = \frac{7}{4}$$

$$\log\left(\frac{1.171}{1.088}\right)^x = \log\left(\frac{7}{4}\right)$$

$$x\log\left(\frac{1.171}{1.088}\right) = \log\left(\frac{7}{4}\right)$$

$$x = \frac{\log(7/4)}{\log(1.171/1.088)}.$$

Checking the answer with a calculator, we get

$$x = \frac{\log(7/4)}{\log(1.171/1.088)} = 7.612,$$

and we see that

$$4(1.171)^{7.612} = 13.302 \qquad 7(1.088)^{7.6} = 13.302.$$

37. Taking natural logs, we get

$$e^{x+5} = 7 \cdot 2^x$$

$$\ln e^{x+5} = \ln(7 \cdot 2^x)$$

$$x + 5 = \ln 7 + \ln 2^x$$

$$x + 5 = \ln 7 + x\ln 2$$

$$x - x\ln 2 = \ln 7 - 5$$

$$x(1 - \ln 2) = \ln 7 - 5$$

$$x = \frac{\ln 7 - 5}{1 - \ln 2}$$

41. Taking logs and using the log rules:

$$\log(ab^x) = \log c$$

$$\log a + \log b^x = \log c$$

$$\log a + x\log b = \log c$$

$$x\log b = \log c - \log a$$

$$x = \frac{\log c - \log a}{\log b}.$$

45.

$$58e^{4t+1} = 30$$

$$e^{4t+1} = \frac{30}{58}$$

$$\ln e^{4t+1} = \ln\left(\frac{30}{58}\right)$$

$$4t + 1 = \ln\left(\frac{30}{58}\right)$$

$$t = \frac{1}{4}\left(\ln\left(\frac{30}{58}\right) - 1\right).$$

49. (a) $\log(10 \cdot 100) = \log 1000 = 3$
$\log 10 + \log 100 = 1 + 2 = 3$

(b) $\log(100 \cdot 1000) = \log 100{,}000 = 5$
$\log 100 + \log 1000 = 2 + 3 = 5$

(c) $\log \dfrac{10}{100} = \log \dfrac{1}{10} = \log 10^{-1} = -1$
$\log 10 - \log 100 = 1 - 2 = -1$

(d) $\log \dfrac{100}{1000} = \log \dfrac{1}{10} = \log 10^{-1} = -1$
$\log 100 - \log 1000 = 2 - 3 = -1$

(e) $\log 10^2 = 2$
$2 \log 10 = 2(1) = 2$

(f) $\log 10^3 = 3$
$3 \log 10 = 3(1) = 3$

Solutions for Section 4.2

Exercises

1. Let t be the doubling time, then the population is $2P_0$ at time t, so

$$2P_0 = P_0 e^{0.2t}$$
$$2 = e^{0.2t}$$
$$0.2t = \ln 2$$
$$t = \frac{\ln 2}{0.2} \approx 3.466.$$

5. The growth factor for Tritium should be $1 - 0.05471 = 0.94529$, since it is decaying by 5.471% per year. Therefore, the decay equation starting with a quantity of a should be:

$$Q = a(0.94529)^t,$$

where Q is quantity remaining and t is time in years. The half life will be the value of t for which Q is $a/2$, or half of the initial quantity a. Thus, we solve the equation for $Q = a/2$:

$$\frac{a}{2} = a(0.94529)^t$$
$$\frac{1}{2} = (0.94529)^t$$
$$\log(1/2) = \log(0.94529)^t$$
$$\log(1/2) = t \log(0.94529)$$
$$t = \frac{\log(1/2)}{\log(0.94529)} = 12.320.$$

So the half-life is about 12.3 years.

9. To convert to the form $Q = ae^{kt}$, we first say that the right sides of the two equations equal each other (since each equals Q), and then we solve for a and k. Thus, we have $ae^{kt} = 2 \cdot 3^t$. At $t = 0$, we can solve for a:

$$ae^{k \cdot 0} = 2 \cdot 3^0$$
$$a \cdot 1 = 2 \cdot 1$$
$$a = 2.$$

Thus, we have $2e^{kt} = 2 \cdot 3^t$, and we solve for k:

$$2e^{kt} = 2 \cdot 3^t$$
$$e^{kt} = 3^t$$
$$\left(e^k\right)^t = 3^t$$
$$e^k = 3$$
$$\ln e^k = \ln 3$$
$$k = \ln 3 \approx 1.099.$$

Therefore, the equation is $Q = 2e^{1.099t}$.

13. The continuous percent growth rate is the value of k in the equation $Q = ae^{kt}$, which is 0.7.

To convert to the form $Q = ab^t$, we first say that the right sides of the two equations equal each other (since each equals Q), and then we solve for a and b. Thus, we have $ab^t = 0.3e^{0.7t}$. At $t = 0$, we can solve for a:

$$ab^0 = 0.3e^{0.7 \cdot 0}$$
$$a \cdot 1 = 0.3 \cdot 1$$
$$a = 0.3.$$

Thus, we have $0.3b^t = 0.3e^{0.7t}$, and we solve for b:

$$0.3b^t = 0.3e^{0.7t}$$
$$b^t = e^{0.7t}$$
$$b^t = \left(e^{0.7}\right)^t$$
$$b = e^{0.7} \approx 2.014.$$

Therefore, the equation is $Q = 0.3 \cdot 2.014^t$.

17. We want $100e^{-0.07t} = 100(e^{-0.07})^t = ab^t$, so we choose $a = 100$ and $b = e^{-0.07} = 0.9324$. The given exponential function is equivalent to the exponential function $y = 100(0.9324)^t$. Since $1 - 0.9324 = 0.0676$, the annual percent decay rate is 6.76% and the continuous percent decay rate per year is 7% per year.

Problems

21. (a) Let $P(t) = P_0b^t$ describe our population at the end of t years. Since P_0 is the initial population, and the population doubles every 15 years, we know that, at the end of 15 years, our population will be $2P_0$. But at the end of 15 years, our population is $P(15) = P_0b^{15}$. Thus

$$P_0b^{15} = 2P_0$$
$$b^{15} = 2$$
$$b = 2^{\frac{1}{15}} \approx 1.04729$$

Since b is our growth factor, the population is, yearly, 104.729% of what it had been the previous year. Thus it is growing by 4.729% per year.

(b) Writing our formula as $P(t) = P_0e^{kt}$, we have $P(15) = P_0e^{15k}$. But we already know that $P(15) = 2P_0$. Therefore,

$$P_0e^{15k} = 2P_0$$
$$e^{15k} = 2$$
$$\ln e^{15k} = \ln 2$$
$$15k \ln e = \ln 2$$
$$15k = \ln 2$$
$$k = \frac{\ln 2}{15} \approx 0.04621.$$

This tells us that we have a continuous annual growth rate of 4.621%.

25. Take logarithms:

$$3^{(4\log x)} = 5$$
$$\log 3^{(4\log x)} = \log 5$$
$$(4\log x)\log 3 = \log 5$$
$$4\log x = \frac{\log 5}{\log 3}$$
$$\log x = \frac{\log 5}{4\log 3}$$
$$x = 10^{(\log 5)/(4\log 3)}.$$

29. Using $\log a + \log b = \log(ab)$, we can rewrite the equation as

$$\log(x(x-1)) = \log 2$$
$$x(x-1) = 2$$
$$x^2 - x - 2 = 0$$
$$(x-2)(x+1) = 0$$
$$x = 2 \text{ or } -1$$

but $x \neq -1$ since $\log x$ is undefined at $x = -1$. Thus $x = 2$.

33. (a) If we let $y =$ the number of cases of sepsis in year t, then we have $y = ae^{kt}$. We find k using the fact that the number of cases doubles in 5 years:

$$y = ae^{kt}$$
$$2a = ae^{k\cdot 5}$$
$$2 = e^{5k}$$
$$\ln 2 = 5k$$
$$k = \frac{\ln 2}{5} = 0.139.$$

The number of cases of sepsis has been growing at a continuous rate of about 13.9% per year.

(b) We have:

$$y = ae^{0.139t}$$
$$3a = ae^{0.139t}$$
$$3 = e^{0.139t}$$
$$\ln 3 = 0.139t$$
$$t = \frac{\ln 3}{0.139} = 7.904.$$

It will take about 7.9 years for the number of cases to triple.

37. (a) Use $o(t)$ to describe the number of owls as a function of time. After 1 year, we see that the number of owls is 103% of 245, or $o(1) = 245(1.03)$. After 2 years, the population is 103% of that number, or $o(2) = (245(1.03)) \cdot 1.03 = 245(1.03)^2$. After t years, it is $o(t) = 245(1.03)^t$.

(b) We will use $h(t)$ to describe the number of hawks as a function of time. Since $h(t)$ doubles every 10 years, we know that its growth factor is constant and so it is an exponential function with a formula of the form $h(t) = ab^t$. In this case the initial population is 63 hawks, so $h(t) = 63b^t$. We are told that the population in 10 years is twice the current population, that is

$$63b^{10} = 126.$$

Thus,

$$b^{10} = 2$$
$$b = 2^{1/10} \approx 1.072.$$

The number of hawks as a function of time is

$$h(t) = 63(2^{1/10})^t = 63 \cdot 2^{t/10} \approx 63 \cdot (1.072)^t.$$

(c) Looking at Figure 4.1 we see that it takes about 34.2 years for the populations to be equal.

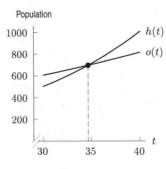

Figure 4.1

41. We have

$$Q = 0.1e^{-(1/2.5)t},$$

and need to find t such that $Q = 0.04$. This gives

$$0.1e^{-\frac{t}{2.5}} = 0.04$$
$$e^{-\frac{t}{2.5}} = 0.4$$
$$\ln e^{-\frac{t}{2.5}} = \ln 0.4$$
$$-\frac{t}{2.5} = \ln 0.4$$
$$t = -2.5 \ln 0.4 \approx 2.291.$$

It takes about 2.3 hours for their BAC to drop to 0.04.

45. Setting the balances equal,

$$\underbrace{4000(1.06)^t}_{\substack{\text{your} \\ \text{balance}}} = \underbrace{3500e^{0.0595t}}_{\substack{\text{your friend's} \\ \text{balance}}}$$

$$\ln(4000(1.06)^t) = \ln(3500e^{0.0595t})$$
$$\ln 4000 + \ln(1.06)^t = \ln 3500 + \ln e^{0.0595t}$$
$$\ln 4000 + t \ln 1.06 = \ln 3500 + 0.0595t$$
$$\ln 4000 - \ln 3500 = 0.0595t - t \ln 1.06 = t(0.0595 - \ln 1.06)$$
$$t = \frac{\ln 4000 - \ln 3500}{0.0595 - \ln 1.06} \approx 108.466$$

Yes, the balances will eventually be equal, but only after 109 years!

49. Population q grows by 3% per hour, compounded hourly, so each day it grows by a factor of $1.03^{24} = 2.033$. Thus, it more than doubles every day, so it grows faster than population m. For any initial populations, in the long term, q has the larger population.

Solutions for Section 4.3

Exercises

1. A is $y = 10^x$, B is $y = e^x$, C is $y = \ln x$, D is $y = \log x$.

5. (a) $10^{-x} \to 0$ as $x \to \infty$.
 (b) The values of $\log x$ get more and more negative as $x \to 0^+$, so

$$\log x \to -\infty.$$

9. See Figure 4.2. The graph of $y = \log(x - 4)$ is the graph of $y = \log x$ shifted to the right 4 units.

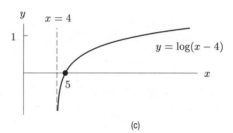

(c)

Figure 4.2

13. We know, by the definition of pH, that $13 = -\log[H^+]$. Therefore, $-13 = \log[H^+]$, and $10^{-13} = [H^+]$. Thus, the hydrogen ion concentration is 10^{-13} moles per liter.

17. We know, by the definition of pH, that $0 = -\log[H^+]$. Therefore, $-0 = \log[H^+]$, and $10^{-0} = [H^+]$. Thus, the hydrogen ion concentration is $10^{-0} = 10^0 = 1$ mole per liter.

Problems

21. This graph could represent exponential decay, so a possible formula is $y = b^x$ with $0 < b < 1$.

25. This graph could represent exponential "growth", with a y-intercept of -1. A possible formula is $y = (-1)b^x = -b^x$ for $b > 1$.

29. Since the domain of $\ln x$ is $x > 0$, the domain of $\ln(x - 3)$ is $(x - 3) > 0$, or $x > 3$.

33. Let I_A and I_B be the intensity of sound A and sound B, respectively. We know that $I_B = 5I_A$ and, by the definition of the decibel rating, we know that $10 \log(I_A/I_0) = 30$. We have:

$$\begin{aligned}
\text{Decibel rating of B} &= 10 \log\left(\frac{I_B}{I_0}\right) \\
&= 10 \log\left(\frac{5I_A}{I_0}\right) \\
&= 10 \log 5 + 10 \log\left(\frac{I_A}{I_0}\right) \\
&= 10 \log 5 + 30 \\
&= 10(0.699) + 30 \\
&\approx 37.
\end{aligned}$$

Notice that although sound B is 5 times as loud as sound A, the decibel rating only goes from 30 to 37.

Solutions for Section 4.4

Exercises

1. This should be graphed on a log scale. Someone who has never been exposed presumably has zero bacteria. Someone who has been slightly exposed has perhaps one thousand bacteria. Someone with a mild case may have ten thousand bacteria, and someone dying of tuberculosis may have hundreds of thousands or millions of bacteria. Using a linear scale, the data points of all the people not dying of the disease would be too close to be readable.

5. **(a)** Using linear regression we find that the linear function $y = 48.097 + 0.803x$ gives a correlation coefficient of $r = 0.9996$. We see from the sketch of the graph of the data that the estimated regression line provides an excellent fit. See Figure 4.3.

   (b) To check the fit of an exponential we make a table of x and $\ln y$:

x	30	85	122	157	255	312
$\ln y$	4.248	4.787	4.977	5.165	5.521	5.704

   Using linear regression, we find $\ln y = 4.295 + 0.0048x$. We see from the sketch of the graph of the data that the estimated regression line fits the data well, but not as well as part (a). See Figure 4.4. Solving for y to put this into exponential form gives

   $$e^{\ln y} = e^{4.295+0.0048x}$$
   $$y = e^{4.295}e^{0.0048x}$$
   $$y = 73.332e^{0.0048x}.$$

   This gives us a correlation coefficient of $r \approx 0.9728$. Note that since $e^{0.0048} = 1.0048$, we could have written $y = 73.332(1.0048)^x$.

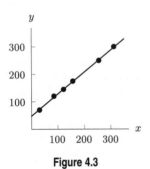

Figure 4.3

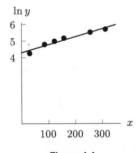

Figure 4.4

   (c) Both fits are good. The linear equation gives a slightly better fit.

Problems

9. **(a)** A log scale is necessary because the numbers are of such different magnitudes. If we used a small scale (such as 0, 10, 20,...) we could see the small numbers but would never get large enough for the big numbers. If we used a large scale (such as counting by 100, 000s), we would not be able to differentiate between the small numbers. In order to see all of the values, we need to use a log scale.

   (b) See Table 4.1.

Table 4.1 *Deaths due to various causes in the US in 1993*

Cause	Log of Deaths	Cause	Log of Deaths
Whooping cough	0.85	Nutritional deficiencies	3.54
Syphilis	1.90	Homicide & legal interv.	4.42
Pregnancy & childbirth	2.48	Motor vehicle	4.62
Meningitis	2.90	Malignant neoplasms	5.72
Viral hepatitis	3.40	All causes	6.36

(c) See Figure 4.5.

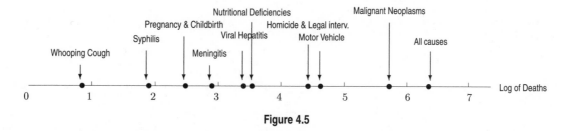

Figure 4.5

13. The figure represents populations using logs, which are exponents of 10. For instance, Greasewood, AZ corresponds to the logarithm 2.3. This means that

$$\text{Population of Greasewood} = 10^{2.3} = 200 \quad \text{after rounding}$$

The population of the other places are in Table 4.2.

Table 4.2 *Approximate populations of eleven different localities*

Locality	Exponent	Approx. population
Lost Springs, Wy	0.6	4
Greasewood, Az	2.3	200
Bar Harbor, Me	3.4	2,500
Abilene, Tx	5.1	130,000
Worcester, Ma	5.6	400,000
Massachusetts	6.8	6,300,000
Chicago	6.9	7,900,000
New York	7.3	20,000,000
California	7.5	32,000,000
US	8.4	250,000,000
World	9.8	6,300,000,000

17.

Table 4.3

x	0	1	2	3	4	5
$y = \ln(3^x)$	0	1.0986	2.1972	3.2958	4.3944	5.4931

Table 4.4

x	0	1	2	3	4	5
$g(x) = \ln(2 \cdot 5^x)$	0.6931	2.3026	3.9120	5.5215	7.1309	8.7403

Yes, the results are linear.

21. (a) Table 4.5 gives values of $L = \ln \ell$ and $W = \ln w$.

Table 4.5 $L = \ln \ell$ and $W = \ln w$ for 16 different fish

Type	1	2	3	4	5	6	7	8
L	2.092	2.208	2.322	2.477	2.501	2.625	2.695	2.754
W	1.841	2.262	2.451	2.918	3.266	3.586	3.691	3.857
Type	9	10	11	12	13	14	15	16
L	2.809	2.874	2.929	2.944	3.025	3.086	3.131	3.157
W	4.184	4.240	4.336	4.413	4.669	4.786	5.131	5.155

These data in Table 4.5 have been plotted in Figure 4.6, and a line of best fit has been drawn in. See part (b).

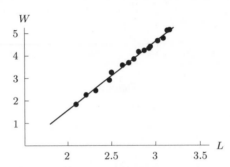

Figure 4.6: Plot of data in Table 4.5 together
with line of best fit

(b) The formula for the line of best fit is $W = 3.06L - 4.54$, as determined using a spreadsheet. However, you could also obtain comparable results by fitting a line by eye.

(c) We have

$$W = 3.06L - 4.54$$
$$\ln w = 3.06 \ln \ell - 4.54$$
$$\ln w = \ln \ell^{3.06} - 4.54$$
$$w = e^{\ln \ell^{3.06} - 4.54}$$
$$= \ell^{3.06} e^{-4.54} \approx 0.011 \ell^{3.06}.$$

(d) Weight tends to be directly proportional to volume, and in many cases volume tends to be proportional to the cube of a linear dimension (e.g., length). Here we see that w is in fact very nearly proportional to the cube of ℓ.

Solutions for Chapter 4 Review

Exercises

1. To convert to the form $Q = ae^{kt}$, we first say that the right sides of the two equations equal each other (since each equals Q), and then we solve for a and k. Thus, we have $ae^{kt} = 12(0.9)^t$. At $t = 0$, we can solve for a:

$$ae^{k \cdot 0} = 12(0.9)^0$$
$$a \cdot 1 = 12 \cdot 1$$
$$a = 12.$$

Thus, we have $12e^{kt} = 12(0.9)^t$, and we solve for k:

$$12e^{kt} = 12(0.9)^t$$
$$e^{kt} = (0.9)^t$$
$$\left(e^k\right)^t = (0.9)^t$$
$$e^k = 0.9$$
$$\ln e^k = \ln 0.9$$
$$k = \ln 0.9 \approx -0.105.$$

Therefore, the equation is $Q = 12e^{-0.105t}$.

5. The continuous percent growth rate is the value of k in the equation $Q = ae^{kt}$, which is -10.

To convert to the form $Q = ab^t$, we first say that the right sides of the two equations equal each other (since each equals Q), and then we solve for a and b. Thus, we have $ab^t = 7e^{-10t}$. At $t = 0$, we can solve for a:

$$ab^0 = 7e^{-10 \cdot 0}$$
$$a \cdot 1 = 7 \cdot 1$$
$$a = 7.$$

Thus, we have $7b^t = 7e^{-10t}$, and we solve for b:

$$7b^t = e^{-10t}$$
$$b^t = e^{-10t}$$
$$b^t = \left(e^{-10}\right)^t$$
$$b = e^{-10} \approx 0.0000454.$$

Therefore, the equation is $Q = 7(0.0000454)^t$.

9. We are solving for an exponent, so we use logarithms. We first divide both sides by 40 and then use logs:

$$40e^{-0.2t} = 12$$
$$e^{-0.2t} = 0.3$$
$$\ln(e^{-0.2t}) = \ln(0.3)$$
$$-0.2t = \ln(0.3)$$
$$t = \frac{\ln(0.3)}{-0.2} = 6.020.$$

13. **(a)** For $f(x) = 10^x$,

$$\text{Domain of } f(x) \text{ is all } x$$

$$\text{Range of } f(x) \text{ is all } y > 0.$$

There is one asymptote, the horizontal line $y = 0$.

(b) Since $g(x) = \log x$ is the inverse function of $f(x) = 10^x$, the domain of $g(x)$ corresponds to range of $f(x)$ and range of $g(x)$ corresponds to domain of $g(x)$.

$$\text{Domain of } g(x) \text{ is all } x > 0$$

$$\text{Range of } g(x) \text{ is all } y.$$

The asymptote of $f(x)$ becomes the asymptote of $g(x)$ under reflection across the line $y = x$. Thus, $g(x)$ has one asymptote, the line $x = 0$.

17. Dividing by 13 and 25 before taking logs gives

$$13e^{0.081t} = 25e^{0.032t}$$

$$\frac{e^{0.081t}}{e^{0.032t}} = \frac{25}{13}$$

$$e^{0.081t-0.032t} = \frac{25}{13}$$

$$\ln e^{0.049t} = \ln\left(\frac{25}{13}\right)$$

$$0.049t = \ln\left(\frac{25}{13}\right)$$

$$t = \frac{1}{0.049}\ln\left(\frac{25}{13}\right) \approx 13.345.$$

21. Using the fact that $A^{-1} = 1/A$ and the log rules:

$$\ln(A+B) - \ln(A^{-1}+B^{-1}) = \ln(A+B) - \ln\left(\frac{1}{A} + \frac{1}{B}\right)$$

$$= \ln(A+B) - \ln\frac{A+B}{AB}$$

$$= \ln\left((A+B) \cdot \frac{AB}{A+B}\right)$$

$$= \ln(AB).$$

Problems

25. (a) Let t be the number of years in a man's age above 30 (i.e. let t =the man's age-30) and let M_0 denote his bone mass at age 30. If he is losing 2% per year, then 98% remains after each year, and thus we can say that $M(t) = M_0(0.98)^t$, where $M(t)$ represents the man's bone mass t years after age 30. But we want a formula describing bone mass in terms of a, his age. Since t is number of years in his age over 30, $t = a - 30$. So, we can substitute $a - 30$ for t in our formula to find an expression in terms of a:

$$M(a) = M_0(0.98)^{(a-30)}.$$

(b) We want to know for what value of a

$$M(a) = \frac{1}{2}M_0$$

Therefore, we will solve $M_0(0.98)^{(a-30)} = \frac{1}{2}M_0$

$$(0.98)^{(a-30)} = \frac{1}{2}$$

$$\log\left((0.98)^{(a-30)}\right) = \log\frac{1}{2} = \log 0.5$$

$$(a - 30)\log(0.98) = \log 0.5$$

$$a - 30 = \frac{\log 0.5}{\log 0.98}$$

$$a = 30 + \frac{\log 0.5}{\log(0.98)} \approx 64.3$$

The average man will have lost half his bone mass at approximately 64.3 years of age.

29. For what value of t will $Q(t) = 0.23Q_0$?

$$0.23Q_0 = Q_0 e^{-0.000121t}$$
$$0.23 = e^{-0.000121t}$$
$$\ln 0.23 = \ln e^{-0.000121t}$$
$$\ln 0.23 = -0.000121t$$
$$t = \frac{\ln 0.23}{-0.000121} = 12146.082.$$

So the skull is about 12,146 years old.

33. (a) Solving for x exactly:

$$\frac{3^x}{5^{(x-1)}} = 2^{(x-1)}$$
$$3^x = 5^{x-1} \cdot 2^{x-1}$$
$$3^x = (5 \cdot 2)^{x-1}$$
$$3^x = 10^{x-1}$$
$$\log 3^x = \log 10^{x-1}$$
$$x \log 3 = (x-1)\log 10 = (x-1)(1)$$
$$x \log 3 = x - 1$$
$$x \log 3 - x = -1$$
$$x(\log 3 - 1) = -1$$
$$x = \frac{-1}{\log 3 - 1} = \frac{1}{1 - \log 3}$$

(b)

$$-3 + e^{x+1} = 2 + e^{x-2}$$
$$e^{x+1} - e^{x-2} = 2 + 3$$
$$e^x e^1 - e^x e^{-2} = 5$$
$$e^x(e^1 - e^{-2}) = 5$$
$$e^x = \frac{5}{e - e^{-2}}$$
$$\ln e^x = \ln \left(\frac{5}{e - e^{-2}} \right)$$
$$x = \ln \left(\frac{5}{e - e^{-2}} \right)$$

(c)

$$\ln(2x - 2) - \ln(x - 1) = \ln x$$
$$\ln \left(\frac{2x - 2}{x - 1} \right) = \ln x$$
$$\frac{2x - 2}{x - 1} = x$$
$$\frac{2(x - 1)}{(x - 1)} = x$$
$$2 = x$$

(d) Let $z = 3^x$, then $z^2 = (3^x)^2 = 9^x$, and so we have

$$z^2 - 7z + 6 = 0$$
$$(z - 6)(z - 1) = 0$$
$$z = 6 \quad \text{or} \quad z = 1.$$

Thus, $3^x = 1$ or $3^x = 6$, and so $x = 0$ or $x = \ln 6/\ln 3$.

(e)

$$\ln\left(\frac{e^{4x}+3}{e}\right) = 1$$

$$e^1 = \frac{e^{4x}+3}{e}$$

$$e^2 = e^{4x}+3$$

$$e^2 - 3 = e^{4x}$$

$$\ln(e^2 - 3) = \ln\left(e^{4x}\right)$$

$$\ln(e^2 - 3) = 4x$$

$$\frac{\ln(e^2 - 3)}{4} = x$$

(f)

$$\frac{\ln(8x) - 2\ln(2x)}{\ln x} = 1$$

$$\ln(8x) - 2\ln(2x) = \ln x$$

$$\ln(8x) - \ln\left((2x)^2\right) = \ln x$$

$$\ln\left(\frac{8x}{(2x)^2}\right) = \ln x$$

$$\ln\left(\frac{8x}{4x^2}\right) = \ln x$$

$$\frac{8x}{4x^2} = x$$

$$8x = 4x^3$$

$$4x^3 - 8x = 0$$

$$4x(x^2 - 2) = 0$$

$$x = 0, \sqrt{2}, -\sqrt{2}$$

Only $\sqrt{2}$ is a valid solution, because when $-\sqrt{2}$ and 0 are substituted into the original equation we are taking the logarithm of negative numbers and 0, which is undefined.

37. (a) Table 4.6 describes the height of the ball after n bounces:

Table 4.6

n	$h(n)$
0	6
1	90% of $6 = 6(0.9) = 5.4$
2	90% of $5.4 = 5.4(0.9) = 6(0.9)(0.9) = 6(0.9)^2$
3	90% of $6(0.9)^2 = 6(0.9)^2 \cdot (0.9) = 6(0.9)^3$
4	$6(0.9)^3 \cdot (0.9) = 6(0.9)^4$
5	$6(0.9)^5$
$\vdots$	$\vdots$
n	$6(0.9)^n$

so $h(n) = 6(0.9)^n$.

(b) We want to find the height when $n = 12$, so we will evaluate $h(12)$:

$$h(12) = 6(0.9)^{12} \approx 1.695 \text{ feet (about 1 ft 8.3 inches).}$$

(c) We are looking for the values of n for which $h(n) \leq 1$ inch $= \frac{1}{12}$ foot. So

$$h(n) \leq \frac{1}{12}$$

$$6(0.9)^n \leq \frac{1}{12}$$

$$(0.9)^n \leq \frac{1}{72}$$

$$\log(0.9)^n \leq \log \frac{1}{72}$$

$$n \log(0.9) \leq \log \frac{1}{72}$$

Using your calculator, you will notice that $\log(0.9)$ is negative. This tells us that when we divide both sides by $\log(0.9)$, we must reverse the inequality. We now have

$$n \geq \frac{\log \frac{1}{72}}{\log(0.9)} \approx 40.591$$

So, the ball will rise less than 1 inch by the 41st bounce.

CHECK YOUR UNDERSTANDING

1. True. If x is a positive number, $\log x$ is defined and $10^{\log x} = x$.

5. True. The value of $\log n$ is the exponent to which 10 is raised to get n.

9. True. The natural log function and the e^x function are inverses.

13. True. Since $y = \log \sqrt{x} = \log(x^{1/2}) = \frac{1}{2} \log x$.

17. True. Divide both sides of the first equation by 50. Then take the log of both sides and finally divide by $\log 0.345$ to solve for t.

21. False. Since $\frac{1}{4} = \frac{1}{2} \cdot \frac{1}{2}$, it takes only two half-life periods. That is 10 hours.

25. True. For example, astronomical distances.

29. False. The fit will not be as good as $y = x^3$ but an exponential function can be found.

CHAPTER FIVE

Solutions for Section 5.1

Exercises

1. **(a)**

x	-1	0	1	2	3
$g(x)$	-3	0	2	1	-1

The graph of $g(x)$ is shifted one unit to the right of $f(x)$.

(b)

x	-3	-2	-1	0	1
$h(x)$	-3	0	2	1	-1

The graph of $h(x)$ is shifted one unit to the left of $f(x)$.

(c)

x	-2	-1	0	1	2
$k(x)$	0	3	5	4	2

The graph $k(x)$ is shifted up three units from $f(x)$.

(d)

x	-1	0	1	2	3
$m(x)$	0	3	5	4	2

The graph $m(x)$ is shifted one unit to the right and three units up from $f(x)$.

5. $m(n) - 3.7 = \dfrac{1}{2}n^2 - 3.7$

Sketch by shifting the graph of $m(n) = \frac{1}{2}n^2$ down by 3.7 units, as in Figure 5.1.

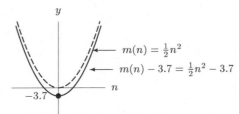

Figure 5.1

9. $m(n+3) + 7 = \dfrac{1}{2}(n+3)^2 + 7 = \left(\dfrac{1}{2}n^2 + 3n + \dfrac{9}{2}\right) + 7 = \dfrac{1}{2}n^2 + 3n + \dfrac{23}{2}$.

To sketch, shift the graph of $m(n) = \frac{1}{2}n^2$ by 3 units to the left and 7 units up, as in Figure 5.2.

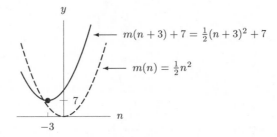

Figure 5.2

13. $k(w) + 1.8 = 3^w + 1.8$

To sketch, shift the graph of $k(w) = 3^w$ up by 1.8 units, as in Figure 5.3.

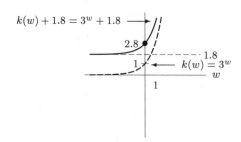

Figure 5.3

17. **(a)** This is the graph of the function $y = |x|$ shifted both up and to the right. Thus the formula is (vi).

(b) This is the graph of the function $y = |x|$ shifted to the right. Thus the formula is (iii).

(c) This is the graph of the function $y = |x|$ shifted down. Thus formula is (ii).

(d) This is the graph of the function $y = |x|$ shifted to the left. Thus the formula is (v).

(e) This is the graph of the function $y = |x|$. Thus the formula is (i).

(f) This is the graph of the function $y = |x|$ shifted up. Thus formula is (iv).

21. Since the $-a$ is an outside change, this transformation shifts the entire graph of $q(z)$ down by a units. That is, for every z, the value of $q(z) - a$ is a units less than $q(z)$.

25. From the inside change, we know that the graph is shifted $2b$ units to the right. From the outside change, we know that it is shifted ab units up. So, for any given z value, the graph of $q(z - 2b) + ab$ is $2b$ units to the right and ab units above the graph of $q(z)$.

Problems

29. **(a)** $f(-6) = ((-6)/2)^3 + 2 = (-3)^3 + 2 = -27 + 2 = -25$

(b) We are trying to find x so that $f(x) = 6$. Setting $f(x) = -6$, we have

$$-6 = \left(\frac{x}{2}\right)^3 + 2$$

$$-8 = \left(\frac{x}{2}\right)^3$$

$$-2 = \left(\frac{x}{2}\right)$$

$$-4 = x.$$

Thus, $f(x) = -6$ for $x = -4$.

(c) In part (a), we found that $f(-6) = -25$. This means that the point $(-6, f(-6))$, or $(-6, -25)$ is on the graph of $f(x)$. We call this point A in Figure 5.4. In part (b), we found that $f(x) = -6$ at $x = -4$. This means the point $(-4, -6)$ is also on the graph of $f(x)$. We call this point B in Figure 5.4.

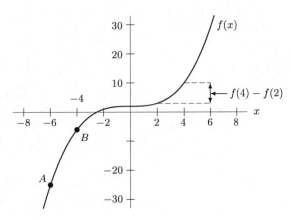

Figure 5.4

(d) We have $f(4) = (4/2)^3 + 2 = 8 + 2 = 10$ and $f(2) = (2/2)^3 + 2 = 1^3 + 2 = 3$. Thus $f(4) - f(2) = 10 - 3 = 7$. This is shown in Fig 5.4.

(e) If $a = -2$, we have $f(a + 4) = f(-2 + 4) = f(2) = 3$. Thus, $f(a + 4) = 3$ for $a = -2$. $f(-2) + 4 = (-2/2)^3 + 2 + 4 = -1 + 2 + 4 = 5$. Thus, $f(a) + 4 = 5$ for $a = -2$.

(f) $f(a+4) = f(-2+4) = f(2)$. Thus, an x-value of 2 corresponds to $f(a+4)$ for $a = -2$. $f(a)+4 = f(-2)+4 = 5$ for $a = -2$. To find an x-value which corresponds to $f(a) + 4$, we need to find the value of x for which $f(x) = 5$. Setting $f(x) = 5$,

$$\left(\frac{x}{2}\right)^3 + 2 = 5$$

$$\frac{x^3}{8} + 2 = 5$$

$$\frac{x^3}{8} = 3$$

$$x^3 = 24$$

$$x = \sqrt[3]{24} = 2\sqrt[3]{3}$$

$$\approx 2.884.$$

33. (a) Notice that the value of $h(x)$ at every value of x is 2 less than the value of $f(x)$ at the same x value. Thus

$$h(x) = f(x) - 2.$$

(b) Observe that $g(0) = f(1)$, $g(1) = f(2)$, and so on. In general,

$$g(x) = f(x + 1).$$

(c) The values of $i(x)$ are two less than the values of $g(x)$ at the same x value. Thus

$$i(x) = f(x + 1) - 2.$$

37. (a) There are many possible graphs, but all should show seasonally-related cycles of temperature increases and decreases, as in Figure 5.5.

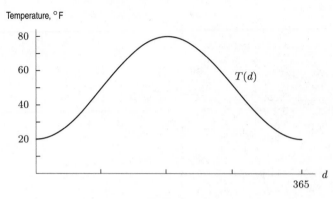

Figure 5.5

(b) While there are a wide variety of correct answers, the value of $T(6)$ is a temperature for a day in early January, $T(100)$ for a day in mid-April, and $T(215)$ for a day in early August. The value for $T(371) = T(365 + 6)$ should be close to that of $T(6)$.

(c) Since there are usually 365 days in a year, $T(d)$ and $T(d + 365)$ represent average temperatures on days which are a year apart.

(d) $T(d + 365)$ is the average temperature on the same day of the year a year earlier. They should be about the same value. Therefore, the graph of $T(d + 365)$ should be about the same as that of $T(d)$.

(e) The graph of $T(d) + 365$ is a shift upward of $T(d)$, by 365 units. It has no significance in practical terms, other than to represent a temperature that is $365°$ hotter than the average temperature on day d.

Solutions for Section 5.2

Exercises

1.

Table 5.1

p	-3	-2	-1	0	1	2	3
$f(p)$	0	-3	-4	-3	0	5	12

Table 5.2

p	-3	-2	-1	0	1	2	3
$g(p)$	12	5	0	-3	-4	-3	0

Table 5.3

p	-3	-2	-1	0	1	2	3
$h(p)$	0	3	4	3	0	-5	-12

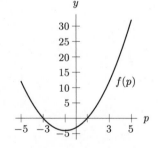 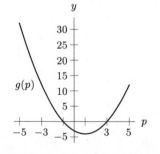 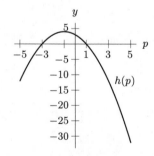

Figure 5.6: Graphs of $f(p)$, $g(p)$, and $h(p)$

Since $g(p) = f(-p)$, the graph of g is a horizontal reflection of the graph of f across the y-axis. Since $h(p) = -f(p)$, the graph of h is a reflection of the graph of f across the p-axis.

5.

$$y = -m(n) = -(n)^2 + 4n - 5$$

To graph this function, reflect the graph of m across the n-axis.

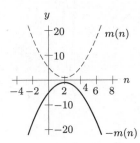

Figure 5.7: $y = -m(n)$

9.

$$y = -m(-n) + 3 = -(-n)^2 + 4(-n) - 5 + 3$$
$$= -n^2 - 4n - 2.$$

To graph this function, first reflect the graph of m across the y-axis, then reflect it across the n-axis, and finally shift it up by 3 units.

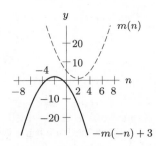

Figure 5.8: $y = -m(-n) + 3$

13.

$$y = -k(-w) = -3^{-w}$$

To graph this function, first reflect the graph of k across the y-axis, then reflect it again across the w-axis.

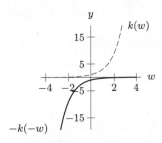

Figure 5.9: $y = -k(-w)$

17.

$$y = -3 - k(w) = -3 - 3^w$$

To graph this function, reflect the graph of k across the the w-axis and then shift it down 3 units.

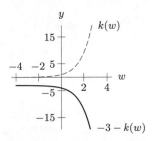

Figure 5.10: $y = -3 - k(w)$

21. Since $f(-x) = (-x)^5 + 3(-x)^3 - 2 = -x^5 - 3x^3 - 2$ is equal to neither $f(x)$ or $-f(x)$, the function is neither even nor odd.

Problems

25. The graphs in Figure 5.11 are reflections of each other across the x-axis. To see this algebraically, note that

$$y = \log\left(\frac{1}{x}\right) = \log 1 - \log x = -\log x.$$

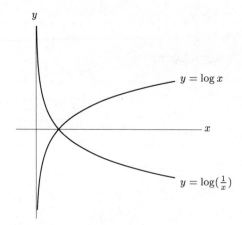

Figure 5.11

29. (a) Figure 5.12 shows the graph of a function f that is symmetric across the y-axis.
    (b) Figure 5.13 shows the graph of function f that is symmetric across the origin.

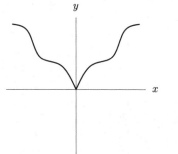

Figure 5.12: The graph of $f(x)$
that is symmetric across the
y-axis

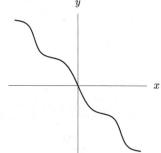

Figure 5.13: The graph of $f(x)$
that is symmetric across the
origin

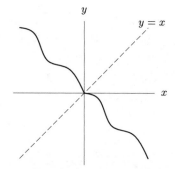

Figure 5.14: The graph of $f(x)$ that is
symmetric across the line $y = x$

(c) Figure 5.14 shows the graph of function f that is symmetric across the line $y = x$.

33. The argument that $f(x)$ is not odd is correct. However, the statement "something is either even or odd" is false. This function is neither an odd function nor an even function.

37. To show that $f(x) = x^{1/3}$ is an odd function, we must show that $f(x) = -f(-x)$:

$$-f(-x) = -(-x)^{1/3} = x^{1/3} = f(x).$$

However, not all power functions are odd. The function $f(x) = x^2$ is an even function because $f(x) = f(-x)$ for all x. Another counter-example is $f(x) = \sqrt{x} = x^{1/2}$. This function is not odd because it is not defined for negative values of x.

Solutions for Section 5.3

Exercises

1. See Figure 5.15.

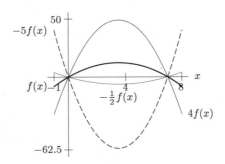

Figure 5.15

5.

Table 5.4

x	-3	-2	-1	0	1	2	3
$f(x)$	-4	-1	2	3	0	-3	-6
$f(-x)$	-6	-3	0	3	2	-1	-4
$-f(x)$	4	1	-2	-3	0	3	6
$f(x) - 2$	-6	-3	0	1	-2	-5	-8
$f(x - 2)$	–	–	-4	-1	2	3	0
$f(x) + 2$	-2	1	4	5	2	-1	-4
$f(x + 2)$	2	3	0	-3	-6	–	–
$2f(x)$	-8	-2	4	6	0	-6	-12
$-f(x)/3$	$4/3$	$1/3$	$-2/3$	-1	0	1	2

9. (i) i: The graph of $y = f(x)$ has been stretched vertically by a factor of 2.
   (ii) c: The graph of $y = f(x)$ has been stretched vertically by 1/3, or compressed.
   (iii) b: The graph of $y = f(x)$ has been reflected over the x-axis and raised by 1.
   (iv) g: The graph of $y = f(x)$ has been shifted left by 2, and raised by 1.
   (v) d: The graph of $y = f(x)$ has been reflected over the y-axis.

13. Since $g(x) = x^2$, $-g(x) = -x^2$. The graph of $g(x)$ is flipped over the x-axis. See Figure 5.16.

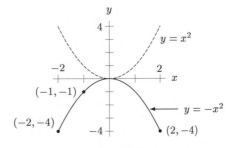

Figure 5.16

Problems

17. **(a) (iii)** The number of gallons needed to cover the house is $f(A)$; two more gallons will be $f(A) + 2$.
    (b) (i) To cover the house twice, you need $f(A) + f(A) = 2f(A)$.
    (c) (ii) The sign is an extra 2 ft^2 so we need to cover the area $A + 2$. Since $f(A)$ is the number of gallons needed to cover A square feet, $f(A + 2)$ is the number of gallons needed to cover $A + 2$ square feet.

21.

(a)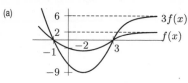

(b)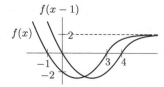

(c)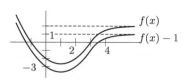

(d)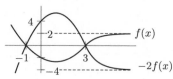

(e)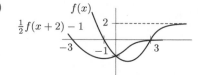

(f)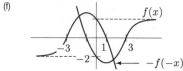

25. **(a)** This figure is the graph of $f(t)$ shifted upwards by two units. Thus its formula is $y = f(t) + 2$. Since on the graph of $f(t)$ the asymptote occurs at $y = 5$ on this graph the asymptote must occur at $y = 7$.
    (b) This figure is the graph of $f(t)$ shifted to the left by one unit. Thus its formula is $y = f(t + 1)$. Since on the graph of $f(t)$ the asymptote occurs at $y = 5$, on this graph the asymptote also occurs at $y = 5$. Note that a horizontal shift does not affect the horizontal asymptotes.
    (c) This figure is the graph of $f(t)$ shifted downwards by three units and to the right by two units. Thus its formula is $y = f(t - 2) - 3$. Since on the graph of $f(t)$ the asymptote occurs at $y = 5$, on this graph the asymptote must occur at $y = 2$. Again, the horizontal shift does not affect the horizontal asymptote. However, outside changes (vertical shifts) do change the horizontal asymptote.

Solutions for Section 5.4

Exercises

1. If $x = -2$, then $f(\frac{1}{2}x) = f(\frac{1}{2}(-2)) = f(-1) = 7$, and if $x = 6$, then $f(\frac{1}{2}x) = f(\frac{1}{2} \cdot 6) = f(3) = 8$. In general, $f(\frac{1}{2}x)$ is defined for values of x which are twice the values for which $f(x)$ is defined.

Table 5.5

x	-6	-4	-2	0	2	4	6
$f(\frac{1}{2}x)$	2	3	7	-1	-3	4	8

5. See Figure 5.17.

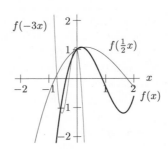

Figure 5.17

Problems

9. (a) (ii)   The $5 tip is added to the fare $f(x)$, so the total is $f(x) + 5$.
   (b) (iv)   There were 5 extra miles so the trip was $x + 5$. I paid $f(x + 5)$.
   (c) (i)   Each trip cost $f(x)$ and I paid for 5 of them, or $5f(x)$.
   (d) (iii)   The miles were 5 times the usual so $5x$ is the distance, and the cost is $f(5x)$.

13. The graph of $y = f(2x)$ is a horizontal compression of the graph of $y = f(x)$ by a factor of 2. The graph of $y = f(-\frac{x}{3}) = f(-\frac{1}{3}x)$ is both a horizontal stretch by a factor of 3 and a flip across the y-axis.

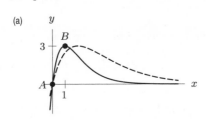 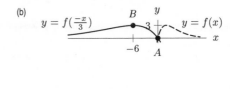

17. (a) Graphing f and g shows that there is a vertical shift of $+1$. See Figure 5.18.

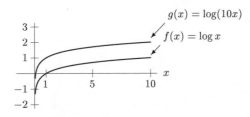

Figure 5.18: A vertical shift of $+1$

(b) Using the property that $\log(ab) = \log a + \log b$, we have

$$g(x) = \log(10x) = \log 10 + \log x = 1 + f(x).$$

Thus, $g(x)$ is $f(x)$ shifted vertically upward by 1.
   (c) Using the same property of logarithms

$$\log(ax) = \log a + \log x \qquad \text{so} \qquad k = \log a.$$

21. If profits are $r(t) = 0.5P(t)$ instead of $P(t)$, then profits are half the dollar level expected. If profits are $s(t) = P(0.5t)$ instead of $P(t)$, then profits are accruing half as fast as the projected rate.

Solutions for Section 5.5

Exercises

1. First, substitute the y-intercept $(0, 5)$ into the standard form of the quadratic function to obtain:

$$5 = a(0)^2 + b(0) + c.$$

This yields $c = 5$. Next because of the symmetry of the parabola, $(-6, 5)$ is also on the graph. See Figure 5.19.

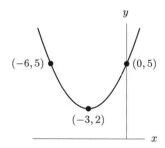

Figure 5.19

The vertex $(-3, 2)$ and the point $(-6, 5)$ when substituted into $y = ax^2 + bx + 5$ give the equations:

$$2 = 9a - 3b + 5$$
$$5 = 36a - 6b + 5.$$

Solving these two linear equations simultaneously yields $a = 1/3$ and $b = 2$. Therefore

$$y = m(x) = \frac{1}{3}x^2 + 2x + 5.$$

5. Since the vertex is $(3, 3)$, we use the form $y = a(x - h)^2 + k$, with $h = 3$ and $k = 3$. We solve for a, substituting in the second point, $(5, 5)$.

$$y = a(x - 3)^2 + 3$$
$$5 = a(5 - 3)^2 + 3$$
$$2 = 4a$$
$$\frac{1}{2} = a.$$

Thus, an equation for the parabola is

$$y = \frac{1}{2}(x - 3)^2 + 3.$$

9. (a) See Figure 5.20. For g, we have $a = 1, b = 0$, and $c = 3$. Its vertex is at $(0, 3)$, and its axis of symmetry is the y-axis, or the line $x = 0$. This function has no zeros.

(b) See Figure 5.21. For f, we have $a = -2, b = 4$, and $c = 16$. The axis of symmetry is the line $x = 1$ and the vertex is at $(1, 18)$. The zeros, or x-intercepts, are at $x = -2$ and $x = 4$. The y-intercept is at $y = 16$.

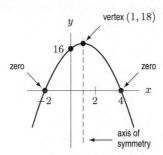

Figure 5.20: $g(x) = x^2 + 3$

Figure 5.21: $f(x) = -2x^2 + 4x + 16$

13. Since the coefficient of x^2 is not 1, we first factor out the coefficient of x^2 from the formula. This gives

$$w(x) = -3\left(x^2 + 10x - \frac{31}{3}\right).$$

We next complete the square of the expression in parentheses. To do this, we add $\left(\frac{1}{2} \cdot 10\right)^2 = 25$ inside the parentheses:

$$w(x) = -3(\underbrace{x^2 + 10x + 25}_{\text{completing the square}} - \underbrace{25}_{\text{compensating term}} - 31/3).$$

Thus,

$$w(x) = -3((x + 5)^2 - 106/3)$$
$$w(x) = -3(x + 5)^2 + 106$$

so the vertex of the graph of this function is $(-5, 106)$, and the axis of symmetry is $x = -5$. Also, since $a = -3$ is negative, the graph is a downward opening parabola.

Problems

17. (a) The graph of g can be found by shifting the graph of f to the right 3 units and then up 2 units; $g(x) = f(x - 3) + 2$.
   (b) Yes, g is a quadratic function. To see this, notice that

$$g(x) = (x - 3)^2 + 2$$
$$= x^2 - 6x + 11.$$

   Thus, g is a quadratic function with parameters $a = 1$, $b = -6$, and $c = 11$.
   (c) Figure 5.22 gives graphs of $f(x) = x^2$ and $g(x) = (x - 3)^2 + 2$. Notice that g's axis of symmetry can be found by shifting f's axis of symmetry to the right 3 units. The vertex of g can be found by shifting f's vertex to the right 3 units and then up 2 units.

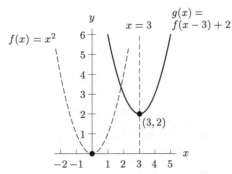

Figure 5.22: The graphs of $f(x) = x^2$ and $g(x) = f(x - 3) + 2$.

21. Yes, we can find the function. Because the vertex is $(1, 4)$, $f(x) = a(x - 1)^2 + 4$ for some a. To find a, we use the fact that $x = -1$ is a zero, that is, the fact that $f(-1) = 0$. We can write $f(-1) = a(-1 - 1)^2 + 4 = 0$, so $4a + 4 = 0$ and $a = -1$. Thus $f(x) = -(x - 1)^2 + 4$.

25. (a)

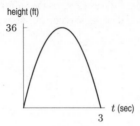

height (ft)

36

3

t (sec)

Figure 5.23

(b) To find t when $d(t) = 0$, either use the graph or factor $-16t^2 + 48t$ and set it equal to zero. Factoring yields $-16t^2 + 48t = -16t(t - 3)$, so $d(t) = 0$ when $t = 0$ or $t = 3$. The first time $d(t) = 0$ is at the moment the tomato is being thrown up into the air. The second time is when the tomato hits the ground.

(c) The maximum height occurs on the axis of symmetry, which is halfway between the zeros, at $t = 1.5$. So, the tomato is highest 1.5 seconds after it is thrown.

(d) The maximum height reached is $d(1.5) = 36$ feet.

29. (a) Factoring gives $h(t) = -16t^2 + 16Tt = 16t(T - t)$. Since $h(t) \geq 0$ only for $0 \leq t \leq T$, the model makes sense only for these values of t.

(b) The times $t = 0$ and $t = T$ give the start and end of the jump. The maximum height occurs halfway in between, at $t = T/2$.

(c) Since $h(t) = 16t(T - t)$, we have

$$h\left(\frac{T}{2}\right) = 16\left(\frac{T}{2}\right)\left(T - \frac{T}{2}\right) = 4T^2.$$

Solutions for Chapter 5 Review

Exercises

1. (a) The input is $2x = 2 \cdot 2 = 4$.
(b) The input is $\frac{1}{2}x = \frac{1}{2} \cdot 2 = 1$.
(c) The input is $x + 3 = 2 + 3 = 5$.
(d) The input is $-x = -2$.

5. $m(-x) = \dfrac{1}{(-x)^2} = \dfrac{1}{x^2} = m(x)$, so $m(x)$ is an even function.

9. $p(-x) = (-x)^2 + 2(-x) = x^2 - 2x$, and $-p(x) = -x^2 - 2x$. Since $p(-x) \neq p(x)$ and $p(-x) \neq -p(x)$, the function p is neither even nor odd.

13. The function has zeros at $x = -1$ and $x = 3$, and appears quadratic, so it could be of the form $y = a(x + 1)(x - 3)$. Since $y = -3$ when $x = 0$, we know that $y = a(0 + 1)(0 - 3) = -3a = -3$, so $a = 1$. Thus $y = (x + 1)(x - 3)$ is a possible formula.

Problems

17. (a) $D(225)$ represents the number of iced cappuccinos sold at a price of $2.25.

(b) $D(p)$ is likely to be a decreasing function. The coffeehouse will probably sell fewer iced cappuccinos if they charge a higher price for them.

(c) p is the price the coffeehouse should charge if they want to sell 180 iced cappuccinos per week.

(d) $D(1.5t)$ represents the number of iced cappuccinos the coffeehouse will sell if they charge one and a half times the average price. $1.5D(t)$ represents 1.5 times the number of cappuccinos sold at the average price. $D(t + 50)$ is the number of iced cappuccinos they will sell if they charge 50 cents more than the average price. $D(t) + 50$ represents 50 more cappuccinos than the number they can sell at the average price.

21. The graph is the cubic function that has been flipped about the x-axis and shifted left and up one unit. Thus, we could try

$$y = -(x + 1)^3 + 1.$$

25. See Figure 5.24.

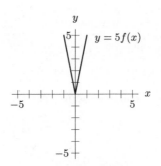

Figure 5.24

29. See Figure 5.25.

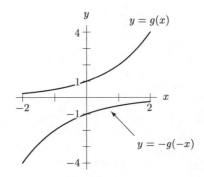

Figure 5.25

33. See Figure 5.26.

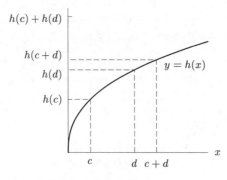

Figure 5.26

37. We have a reflection through the x-axis and a horizontal shift to the right by 1.

$$y = -h(x-1)$$

41. **(a)** We have

$$d_1 = f(30) - f(20) = 650$$
$$d_2 = f(40) - f(30) = 550$$
$$d_3 = f(50) - f(40) = 500$$

(b) d_1, d_2, and d_3 tell us how much building an additional 10 chairs will cost if the carpenter has already built 20, 30, and 40 chairs respectively.

45. **(a)** Sketch varies – should show appropriate seasonal increases/decreases, such as in Figure 5.27.

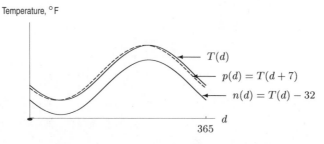

Figure 5.27: Possible graphs of T and n and p

(b) Freezing is $32°$F. If $T(d)$ is the temperature for a particular day, you can determine how far above (or below) freezing $T(d)$ is by subtracting 32 from it. So $n(d) = T(d) - 32$. The sketch of n is the sketch of T shifted 32 units ($32°$F) downward. See Figure 5.27.

(c) Since low temperatures this year are a week ahead of those of last year, the low temperature on the 100th day of this year, $p(100)$, is the same as the low temperature on the 107th day of last year, $T(107)$. More generally, $p(d) = T(d + 7)$. The graph of p is 7 units (7 days) to the left of the graph of T because all low temperatures are occurring seven days earlier. See Figure 5.27.

CHECK YOUR UNDERSTANDING

1. True. The graph of $g(x)$ is a copy of the graph of f shifted vertically up by three units.

5. True. The reflection across the x-axis of $y = f(x)$ is $y = -f(x)$.

9. True. Any point (x, y) on the graph of $y = f(x)$ reflects across the y-axis to the point $(-x, y)$, which lies on the graph of $y = f(-x)$.

13. False. If $f(x) = x^2$, then $f(x + 1) = x^2 + 2x + 1 \neq x^2 + 1 = f(x) + 1$.

17. True.

21. False. Consider $f(x) = x^2$. Shifting up first and then compressing vertically gives the graph of $g(x) = \frac{1}{2}(x^2 + 1) = \frac{1}{2}x^2 + \frac{1}{2}$. Compressing first and then shifting gives the graph of $h(x) = \frac{1}{2}x^2 + 1$.

25. True. This is the definition of a vertex.

29. True. Transform $y = ax^2 + bx + c$ to the form $y = a(x - h)^2 + k$ where it can be seen that if $a < 0$, then the value of y has a maximum at the vertex (h, k), and the parabola opens downward.

Solutions to Tools for Chapter 5

1. $x^2 + 8x = x^2 + 8x + 16 - 16 = (x + 4)^2 - 16$

5. $s^2 + 6s - 8 = s^2 + 6s + 9 - 9 - 8 = (s + 3)^2 - 17$

9. We add and subtract the square of half the coefficient of the c-term, $(\frac{3}{2})^2 = \frac{9}{4}$, to get

$$
\begin{aligned}
c^2 + 3c - 7 &= c^2 + 3c + \frac{9}{4} - \frac{9}{4} - 7 \\
&= \left(c^2 + 3c + \frac{9}{4}\right) - \frac{9}{4} - 7 \\
&= \left(c + \frac{3}{2}\right)^2 - \frac{37}{4}.
\end{aligned}
$$

13. Completing the square yields

$$
x^2 - 2x - 3 = (x^2 - 2x + 1) - 1 - 3 = (x - 1)^2 - 4.
$$

17. Complete the square and write in vertex form.

$$
\begin{aligned}
y &= x^2 + 6x + 3 \\
&= x^2 + 6x + 9 - 9 + 3 \\
&= (x + 3)^2 - 6.
\end{aligned}
$$

The vertex is $(-3, -6)$.

21. Complete the square and write in vertex form.

$$
\begin{aligned}
y &= -x^2 + x - 6 \\
&= -(x^2 - x + 6) \\
&= -\left(x^2 - x + \frac{1}{4} - \frac{1}{4} + 6\right) \\
&= -\left(\left(x - \frac{1}{2}\right)^2 - \frac{1}{4} + 6\right) \\
&= -\left(x - \frac{1}{2}\right)^2 - \frac{23}{4}.
\end{aligned}
$$

The vertex is $(1/2, -23/4)$.

25. Complete the square and write in vertex form.

$$
\begin{aligned}
y &= 2x^2 - 7x + 3 \\
&= 2\left(x^2 - \frac{7}{2}x + \frac{3}{2}\right) \\
&= 2\left(x^2 - \frac{7}{2}x + \frac{49}{16} - \frac{49}{16} + \frac{3}{2}\right) \\
&= 2\left(\left(x - \frac{7}{4}\right)^2 - \frac{49}{16} + \frac{3}{2}\right) \\
&= 2\left(x - \frac{7}{4}\right)^2 - \frac{25}{8}.
\end{aligned}
$$

The vertex is $(7/4, -25/8)$.

29. Complete the square using $(-2/2)^2 = 1$, take the square root of both sides and solve for p.

$$
\begin{aligned}
p^2 - 2p &= 6 \\
p^2 - 2p + 1 &= 6 + 1 \\
(p - 1)^2 &= 7 \\
p - 1 &= \pm\sqrt{7} \\
p &= 1 \pm \sqrt{7}.
\end{aligned}
$$

33. Get the variables on the left side, the constants on the right side and complete the square using $\left(\frac{5}{2}\right)^2 = \frac{25}{4}$.

$$
\begin{aligned}
2s^2 + 10s &= 1 \\
2\left(s^2 + 5s\right) &= 1 \\
2\left(s^2 + 5s + \frac{25}{4}\right) &= 2\left(\frac{25}{4}\right) + 1 \\
2\left(s + \frac{5}{2}\right)^2 &= \frac{25}{2} + 1 \\
2\left(s + \frac{5}{2}\right)^2 &= \frac{27}{2}.
\end{aligned}
$$

Divide by 2, take the square root of both sides and solve for s.

$$
\begin{aligned}
\left(s + \frac{5}{2}\right)^2 &= \frac{27}{4} \\
s + \frac{5}{2} &= \pm\sqrt{\frac{27}{4}} \\
s + \frac{5}{2} &= \pm\frac{\sqrt{27}}{2} \\
s &= -\frac{5}{2} \pm \frac{\sqrt{27}}{2}.
\end{aligned}
$$

37. With $a = 1, b = -4$, and $c = -12$, we use the quadratic formula,

$$
\begin{aligned}
n &= \frac{-b \pm \sqrt{b^2 - 4ac}}{2a} \\
&= \frac{4 \pm \sqrt{(-4)^2 - 4 \cdot 1 \cdot (-12)}}{2 \cdot 1}
\end{aligned}
$$

$$= \frac{4 \pm \sqrt{16 + 48}}{2}$$

$$= \frac{4 \pm \sqrt{64}}{2}$$

$$= \frac{4 \pm 8}{2}.$$

So, $n = 6$ or $n = -2$.

41. Set the equation equal to zero, $z^2 + 4z - 6 = 0$. With $a = 1$, $b = 4$, and $c = -6$, we use the quadratic formula,

$$z = \frac{-b \pm \sqrt{b^2 - 4ac}}{2a}$$

$$= \frac{-4 \pm \sqrt{4^2 - 4 \cdot 1 \cdot (-6)}}{2 \cdot 1}$$

$$= \frac{-4 \pm \sqrt{16 + 24}}{2}$$

$$= \frac{-4 \pm \sqrt{40}}{2}$$

$$= \frac{-4 \pm 2\sqrt{10}}{2}$$

$$= -2 \pm \sqrt{10}.$$

45. Rewrite the equation to equal zero, and factor.

$$n^2 + 4n - 5 = 0$$

$$(n + 5)(n - 1) = 0.$$

So, $n + 5 = 0$ or $n - 1 = 0$, thus $n = -5$ or $n = 1$.

49. Set the equation equal to zero, and use the quadratic formula with $a = 25$, $b = -30$, and $c = 4$.

$$u = \frac{30 \pm \sqrt{(-30)^2 - 4 \cdot 25 \cdot 4}}{2 \cdot 25}$$

$$= \frac{30 \pm \sqrt{900 - 400}}{50}$$

$$= \frac{30 \pm \sqrt{500}}{50}$$

$$= \frac{30 \pm 10\sqrt{5}}{50}$$

$$= \frac{3 \pm \sqrt{5}}{5}.$$

53. Set the equation equal to zero and use factoring.

$$2w^3 - 6w^2 - 8w + 24 = 0$$

$$w^3 - 3w^2 - 4w + 12 = 0$$

$$w^2(w - 3) - 4(w - 3) = 0$$

$$(w - 3)(w^2 - 4) = 0$$

$$(w - 3)(w - 2)(w + 2) = 0.$$

So, $w - 3 = 0$ or $w - 2 = 0$ or $w + 2 = 0$ thus, $w = 3$ or $w = 2$ or $w = -2$.

CHAPTER SIX

Solutions for Section 6.1

Exercises

1. This could be a periodic function with a period of 3. The values of $f(t)$ repeat each time t increases by 3.

5. This does not appear to be a periodic function, because, while it rises and falls, it does not rise and fall to the same level regularly.

9. The period appears to be 3.

Problems

13. The wheel will complete two full revolutions after 20 minutes, so the function is graphed on the interval $0 \leq t \leq 20$. See Figure 6.1.

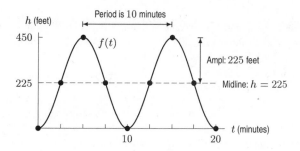

Figure 6.1: Graph of $h = f(t)$, $0 \leq t \leq 20$

17. See Figure 6.2.

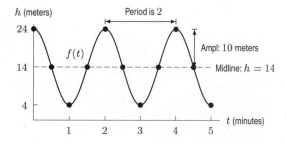

Figure 6.2: Graph of $h = f(t)$, $0 \leq t \leq 5$

21. At $t = 0$, we see $h = 20$, so you are level with the center of the wheel. Your initial position is at three o'clock (or nine o'clock) and initially you are rising. On the interval $0 \le t \le 7$ the wheel completes seven fourths of a revolution. Therefore, if p is the period, we know that

$$\frac{7}{4}p = 7$$

which gives $p = 4$. This means that the ferris wheel takes 4 minutes to complete one full revolution. The minimum value of the function is $h = 5$, which means that you get on and get off of the wheel from a 5 meter platform. The maximum height above the midline is 15 meters, so the wheel's diameter is 30 meters. Notice that the wheel completes a total 2.75 cycles. Since each period is 4 minutes long, you ride the wheel for $4(2.75) = 11$ minutes.

25.

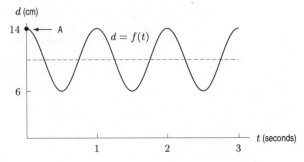

Figure 6.3: Graph of $d = f(t)$ for $0 \le t \le 3$

Since the weight is released at $d = 14$ cm when $t = 0$, it is initially at the point in Figure 6.3 labeled A. The weight will begin to oscillate in the same fashion as described by Figures 6.9 and 6.10. Thus, the period, amplitude, and midline for Figure 6.3 are the same as for Figures 6.9 and 6.10 in the text.

29. By plotting the data in Figure 6.4, we can see that the midline is at $h = 2$ (approximately). Since the maximum value is 3 and the minimum value is 1, we have

$$\text{Amplitude} = 2 - 1 = 1.$$

Finally, we can see from the graph that one cycle has been completed from time $t = 0$ to time $t = 1$, so the period is 1 second.

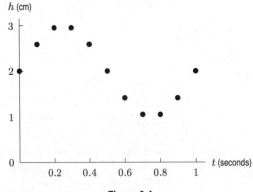

Figure 6.4

Solutions for Section 6.2

Exercises

1.

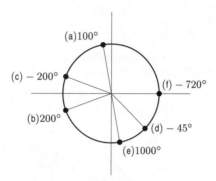

Figure 6.5

(a) $(\cos 100°, \sin 100°) = (-0.174, 0.985)$
(b) $(\cos 200°, \sin 200°) = (-0.940, -0.342)$
(c) $(\cos(-200°), \sin(-200°)) = (-0.940, 0.342)$
(d) $(\cos(-45°), \sin(-45°)) = (0.707, -0.707)$
(e) $(\cos 1000°, \sin 1000°) = (0.174, -0.985)$
(f) $(\cos 720°, \sin 720°) = (1, 0)$

5.

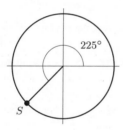 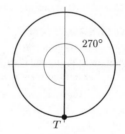 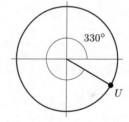

Figure 6.6

$$S = (-0.707, -0.707),\ T = (0, -1),\ U = (0.866, -0.5)$$

9. Since $x = r\cos\theta$ and $y = r\sin\theta$ we have

$$S = (5\cos 225°, 5\sin 225°) = (-3.536, -3.536)$$
$$T = (5\cos 270°, 5\sin 270°) = (0, -5)$$
$$U = (5\cos 330°, 5\sin 330°) = (4.330, -2.5)$$

13. Since the x-coordinate is $r\cos\theta$ and the y-coordinate is $r\sin\theta$ and $r = 3.8$ and $\theta = -180°$, the point is $(3.8\cos(-180), 3.8\sin(-180)) = (-3.8, 0)$.

17. Since the x-coordinate is $r\cos\theta$ and the y-coordinate is $r\sin\theta$ and $r = 3.8$ and $\theta = 1426°$, the point is $(3.8\cos 1426, 3.8\sin 1426) = (3.687, -0.919)$.

21. Since the x-coordinate is $r\cos\theta$ and the y-coordinate is $r\sin\theta$ and $r = 3.8$, the point is $(3.8\cos 225, 3.8\sin 225) = (-3.8\sqrt{2}/2, -3.8\sqrt{2}/2) = (-2.687, -2.687)$.

Problems

25. The car on the ferris wheel starts at the 3 o'clock position. Let's suppose that you see the wheel rotating counterclockwise. (If not, move to the other side of the wheel.)

    The angle $\phi = 420°$ indicates a counterclockwise rotation of the ferris wheel from the 3 o'clock position all the way around once ($360°$), and then two-thirds of the way back up to the top (an additional $60°$). This leaves you in the 1 o'clock position, or at the angle $60°$.

    A negative angle represents a rotation in the opposite direction, that is clockwise. The angle $\theta = -150°$ indicates a rotation from the 3 o'clock position in the clockwise direction, past the 6 o'clock position and two-thirds of the way up to the 9 o'clock position. This leaves you in the 8 o'clock position, or at the angle $210°$. (See Figure 6.7.)

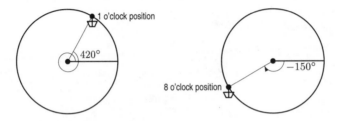

Figure 6.7: The positions and displacements on the ferris wheel described by $420°$ and $-150°$

29. (a) Since the four panels divide a full rotation or $360°$ into four equal spaces, the angle between two adjacent panels is

$$\frac{360°}{4} = 90°.$$

  (b) The angle created by rotating a panel from B to A is equal to the angle between each panel, or $90°$.
  (c) Point B is directly across from point D. So the angle between the two is $180°$.
  (d) If the door moves from B to D, the angle of rotation is $180°$.
  (e) Each person, whether entering or leaving, must rotate the door by $180°$. Thus the total rotation is $(3 + 5)(180°) = 8(180°) = 1440°$. Since $1440° = 4(360°)$ the rotation is equivalent to $0°$. Thus, the panel at point A ends up at point A.

33. Given the angle θ, draw a line l through the origin making an angle θ with the x-axis. Go counterclockwise if $\theta > 0$ and clockwise if $\theta < 0$, wrapping around the unit circle more than once if necessary. Let $P = (x, y)$ be the point where l intercepts the unit circle. Then the definition of sine is that $\sin \theta = y$.

Solutions for Section 6.3

Exercises

1. To convert $45°$ to radians, multiply by $\pi/180°$:

$$45° \left(\frac{\pi}{180°} \right) = \left(\frac{45°}{180°} \right) \pi = \frac{\pi}{4}.$$

Thus we say that the radian measure of a $45°$ angle is $\pi/4$.

5. In order to change from degrees to radians, we multiply the number of degrees by $\pi/180$, so we have $120 \cdot \pi/180$, giving $\frac{2}{3}\pi$ radians.

9. In order to change from radians to degrees, we multiply the number of radians by $180/\pi$, so we have $5\pi \cdot 180/\pi$, giving 900 degrees.

13. In order to change from radians to degrees, we multiply the number of radians by $180/\pi$, so we have $45 \cdot 180/\pi$, giving $8100/\pi \approx 2578.310$ degrees.

17. If we go around 0.75 times, we make three-fourths of a full circle, which is $2\pi \cdot \frac{3}{4} = 3\pi/2$ radians.

21. The arc length, s, corresponding to an angle of θ radians in a circle of radius r is $s = r\theta$. In order to change from degrees to radians, we multiply the number of degrees by $\pi/180$, so we have $-180 \cdot \pi/180$, giving $-\pi$ radians. The negative sign indicates rotation in a clockwise, rather than counterclockwise, direction. Since length cannot be negative, we find the arc length corresponding to π radians. Thus, our arc length is $6.2\pi \approx 19.478$.

Problems

25. Using $s = r\theta$, we have $s = 8(2) = 16$ inches.

29. (a) Negative
   (b) Negative
   (c) Positive
   (d) Positive

33. $\sin \theta = 0.8$, $\cos \theta = -0.6$.

37. As the bob moves from one side to the other, as in Figure 6.8, the string moves through an angle on $10°$. We are therefore looking for the arc length on a circle of radius 3 feet cut off by an angle of $10°$. First we convert $10°$ to radians

$$10° = 10 \cdot \frac{\pi}{180} = \frac{\pi}{18} \text{ radians.}$$

Then we find

$$\text{arc length} = \text{radius} \cdot \text{angle spanned in radians}$$
$$= 3\left(\frac{\pi}{18}\right)$$
$$= \frac{\pi}{6} \text{ feet.}$$

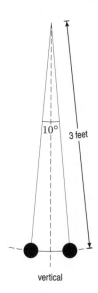

Figure 6.8

41. (a) 1 radian is $180/\pi$ degrees so 30 radians is

$$30 \cdot \frac{180°}{\pi} \approx 1718.873°.$$

To check this answer, divide $1718.873°$ by $360°$ to find this is roughly 5 revolutions. A revolution in radians has a measure of $2\pi \approx 6$, so $5 \cdot 6 = 30$ radians makes sense.

(b) 1 degree is $\pi/180$ radians, so $\pi/6$ degrees is

$$\frac{\pi}{6} \cdot \frac{\pi}{180} = \frac{\pi^2}{6 \cdot 180} \approx 0.00914 \text{ radians.}$$

This makes sense because $\pi/6$ is about $1/2$, and $1/2$ a degree is very small. One radian is about $60°$ so $\frac{1}{2}°$ is a very small part of a radian.

Solutions for Section 6.4

Exercises

1.

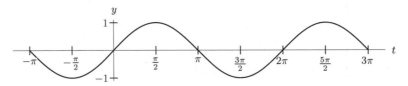

Figure 6.9

(a) (i) For $0 < t < \pi$ and $2\pi < t < 3\pi$ the function $\sin t$ is positive.

(ii) It is increasing for $-\frac{\pi}{2} < t < \frac{\pi}{2}$ and $\frac{3\pi}{2} < t < \frac{5\pi}{2}$.

(iii) For $-\pi < t < 0$ and $\pi < t < 2\pi$ it is concave up.

(b) The function appears to have the maximum rate of increase at $t = 0, 2\pi$.

5. Since the maximum value of the function is 3 and the minimum is 1, its midline is 2, and its amplitude is 1.

9. Since the middle of the clock's face is at 223 cm, the midline is 223 cm, and since the hand is 20 cm long, the amplitude is 20 cm.

13. Since we know that the y-coordinate on the unit circle at $-\pi/3$ is the negative of the y-coordinate at $\pi/3$, and since $\pi/3$ radians is the same as $60°$, we know that $\sin(-\pi/3) = -\sin(\pi/3) = -\sin 60° = -\sqrt{3}/2$.

Problems

17. (a) $7\pi/8$
(b) $\pi - 1$

21. We can sketch these graphs using a calculator or computer. Figure 6.10 gives a graph of $y = \sin\theta$, together with the graphs of $y = 0.5\sin\theta$ and $y = -2\sin\theta$, where θ is in radians and $0 \le \theta \le 2\pi$.

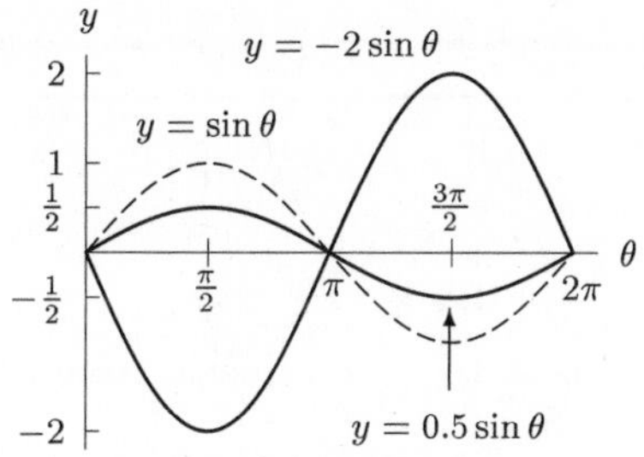

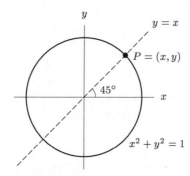

Figure 6.10

These graphs are similar but not the same. The amplitude of $y = 0.5 \sin \theta$ is 0.5 and the amplitude of $y = -2 \sin \theta$ is 2. The graph of $y = -2 \sin \theta$ is vertically reflected relative to the other two graphs. These observations are consistent with the fact that the constant A in the equation

$$y = A \sin \theta$$

may result in a vertical stretching or shrinking and/or a reflection over the x-axis. Note that all three graphs have a period of 2π.

25. Since $45°$ is half $90°$, the point P in Figure 6.11 lies on the line $y = x$. Substituting $y = x$ into the equation of the circle, $x^2 + y^2 = 1$, gives $x^2 + x^2 = 1$. Solving for x, we get

$$2x^2 = 1$$
$$x^2 = \frac{1}{2}$$
$$x = \pm\sqrt{\frac{1}{2}} = \pm\frac{1}{\sqrt{2}}.$$

Since P is in the first quadrant, x and y are positive, so

$$x = \cos 45° = \frac{1}{\sqrt{2}} \qquad \text{and} \qquad y = \sin 45° = \frac{1}{\sqrt{2}}.$$

Figure 6.11: Calculating $\cos 45°$ and $\sin 45°$

29. (a) Slope, m, of segment joining S and T:

$$m = \frac{\sin(a+h) - \sin a}{(a+h) - a} = \frac{\sin(a+h) - \sin a}{h}$$

(b) If $a = 1.7$ and $h = 0.05$,

$$m = \frac{\sin 1.75 - \sin 1.7}{1.75 - 1.7} \approx -0.15$$

Solutions for Section 6.5

Exercises

1. The midline is -8. The amplitude is 7. The period is $2\pi/4 = \pi/2$.

5. We see that the phase shift is -4, since the function is in a form that shows it. To find the horizontal shift, we factor out a 3 within the cosine function, giving us

$$y = 2\cos\left(3\left(t + \frac{4}{3}\right)\right) - 5.$$

Thus, the horizontal shift is $-4/3$.

9. See Figure 6.12.

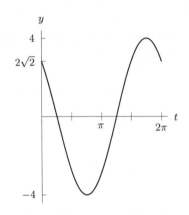

Figure 6.12: $y = 4\cos(t + \pi/4)$

13. This function resembles a cosine curve in that it attains its maximum value when $t = 0$. We know that the smallest value it attains is -3 and that its midline is $y = 0$. Thus its amplitude is 3. It has a period of 4. Thus in the equation

$$f(t) = A\cos(Bt)$$

we know that $A = 3$ and

$$4 = \text{period} = \frac{2\pi}{B}.$$

So $B = \pi/2$, and then

$$f(t) = 3\cos\left(\frac{\pi}{2}t\right).$$

17. The graph is a horizontally and vertically compressed sine function. The midline is $y = 0$. The amplitude is 0.8. We see that $\pi/7 = $ two periods, so the period is $\pi/14$. Hence $B = 2\pi/(\text{period}) = 28$, and so

$$y = 0.8\sin(28\theta).$$

Problems

21. Because the period of $\sin x$ is 2π, and the period of $\sin 2x$ is π, so from the figure in the problem we see that

$$f(x) = \sin x.$$

The points on the graph are $a = \pi/2$, $b = \pi$, $c = 3\pi/2$, $d = 2\pi$, and $e = 1$.

25. $f(t) = 14 + 10\sin\left(\pi t + \dfrac{\pi}{2}\right)$

29. (a) The ferris wheel makes one full revolution in 30 minutes. Since one revolution is $360°$, the wheel turns

$$\frac{360}{30} = 12° \text{ per minute.}$$

(b) The angle representing your position, measured from the 6 o'clock position, is $12t°$. However, the angle shown in Figure 6.13 is measured from the 3 o'clock position, so

$$\theta = (12t - 90)°.$$

(c) With y as shown in Figure 6.13, we have

$$\text{Height} = 225 + y = 225 + 225\sin\theta,$$

so

$$f(t) = 225 + 225\sin(12t - 90)°.$$

Note that the expression $\sin(12t - 90)°$ means $\sin\left((12t - 90)°\right)$.

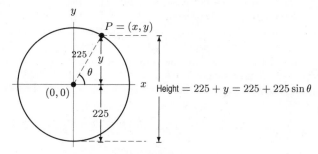

Figure 6.13

(d) Using a calculator, we obtain the graph of $h = f(t) = 225 + 225\sin(12t - 90)°$ in Figure 6.14. The period is 30 minutes, the midline and amplitude are 225 feet.

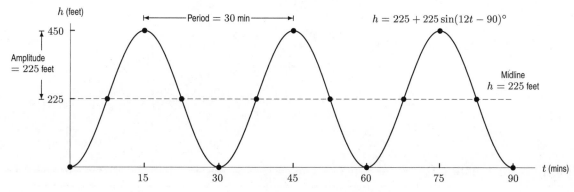

Figure 6.14: On the ferris wheel: Height, h, above ground as function of time, t

33. This function has an amplitude of 3 and a period 1, and resembles a sine graph. Thus $y = 3f(x)$.

37.

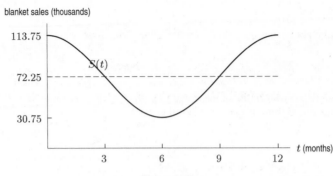

blanket sales (thousands)

Figure 6.15

The amplitude of this graph is 41.5. The period is $P = 2\pi/B = (2\pi)/(\pi/6) = 12$ months. The amplitude of 41.5 tells us that during winter months sales of electric blankets are 41,500 above the average. Similarly, sales reach a minimum of 41,500 below average in the summer months. The period of one year indicates that this seasonal sales pattern repeats annually.

41. (a) Although the graph has a rough wavelike pattern, the wave is not perfectly regular in each 7-day interval. A true periodic function has a graph which is absolutely regular, with values that repeat exactly every period.

(b) Usage spikes every 7 days or so, usually about midweek (8/7, 8/14, 8/21, etc.). It drops to a low point every 7 days or so, usually on Saturday or Sunday (8/10, 8/17, 8/25, etc.). This indicates that scientists use the site less frequently on weekends and more frequently during the week.

(c) See Figure 6.16 for one possible approximation. The function shown here is given by $n = a\cos(B(t-h)) + k$ where t is the number of days from Monday, August 5, and $a = 45,000$, $B = 2\pi/7$, $h = 2$, and $k = 100,000$. The midline $k = 100,000$ tells us that usage rises and falls around an approximate average of 100,000 connections per day. The amplitude $a = 45,000$ tells us that usage tends to rise or fall by about 45,000 from the average over the course of the week. The period is 7 days, or one week, giving $B = 2\pi/7$, and the curve resembles a cosine function shifted to the right by about $h = 2$ days. Thus,

$$n = 45,000\cos\left(\frac{2\pi}{7}(t-2)\right) + 100,000.$$

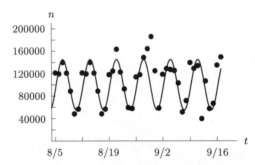

Figure 6.16: Fitting a trigonometric function to the arXiv.org usage data

Solutions for Section 6.6

Exercises

1. $\sin 0° = 0$, $\cos 0° = 1$, $\tan 0° = \sin 0°/\cos 0° = 0/1 = 0$.

5. Since $135°$ is in the second quadrant,
$$\tan 135° = -\tan 45° = -1.$$

9. $-\sqrt{3}$

13. Since $\csc(5\pi/4) = 1/\sin(5\pi/4)$, we know that $\csc(5\pi/4) = 1/(-1/\sqrt{2}) = -\sqrt{2}$.

Problems

17. This looks like a tangent graph. At $\pi/4$, $\tan\theta = 1$. Since on this graph, $f(\pi/2) = 1$, and since it appears to have the same period as $\tan\theta$ without a vertical shift, but shifted $\pi/4$ to the right, a possible formula is $f(\theta) = \tan(\theta - \pi/4)$.

21. Since $\cos\theta = 1/\sec\theta$, $\cos\theta = 1/17$. Using the Pythagorean Identity, $\cos^2\theta + \sin^2\theta = 1$, we have

$$\sin^2\theta + \left(\frac{1}{17}\right)^2 = 1$$

$$\sin^2\theta = 1 - \frac{1}{17^2}$$

$$\sin\theta = \pm\sqrt{\frac{288}{289}} = \pm\frac{\sqrt{288}}{17}.$$

Since $0 \le \theta \le \pi/2$, we know that $\sin\theta \ge 0$, so $\sin\theta = \sqrt{288}/17$.
Using the identity $\tan\theta = \sin\theta/\cos\theta$, we see that $\tan\theta = (\sqrt{288}/17)/(1/17) = \sqrt{288}$.

25. Divide both sides of $\cos^2\theta + \sin^2\theta = 1$ by $\sin^2\theta$. For $\sin\theta \ne 0$,

$$\frac{\cos^2\theta}{\sin^2\theta} + \frac{\sin^2\theta}{\sin^2\theta} = \frac{1}{\sin^2\theta}$$

$$\left(\frac{\cos\theta}{\sin\theta}\right)^2 + 1 = \left(\frac{1}{\sin\theta}\right)^2$$

$$\cot^2\theta + 1 = \csc^2\theta.$$

29. First notice that $\tan\theta = \frac{x}{9}$ so $\tan\theta = \sin\theta/\cos\theta = x/9$, so $\sin\theta = x/9 \cdot \cos\theta$. Now to find $\cos\theta$ by using $1 = \sin^2\theta + \cos^2\theta = (x^2/81)\cos^2\theta + \cos^2\theta = \cos^2\theta(x^2/81+1)$, so $\cos^2\theta = 81/(x^2+81)$ and $\cos\theta = 9/\sqrt{x^2+81}$. Thus, $\sin\theta = (x/9) \cdot (9/\sqrt{x^2+81}) = x/\sqrt{x^2+81}$.

33.

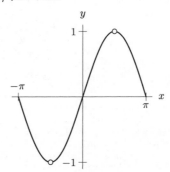

Figure 6.17

Though the function $y = f(x) = \cos x \cdot \tan x$ can be simplified by

$$\cos x \cdot \tan x = \cos x \cdot \frac{\sin x}{\cos x} = \sin x,$$

it is important to notice that $f(x)$ is not defined at the points where $\cos x = 0$. There would be division by zero at such points. Note the holes in the graph which denote undefined values of the function.

Solutions for Section 6.7

Exercises

1. **(a)** Tracing along the graph in Figure 6.18, we see that the approximations for the two solutions are

$$t_1 \approx 1.88 \quad \text{and} \quad t_2 \approx 4.41.$$

Note that the first solution, $t_1 \approx 1.88$, is in the second quadrant and the second solution, $t_2 \approx 4.41$, is in the third quadrant. We know that the cosine function is negative in those two quadrants. You can check the two solutions by substituting them into the equation:

$$\cos 1.88 \approx -0.304 \quad \text{and} \quad \cos 4.41 \approx -0.298,$$

both of which are close to -0.3.

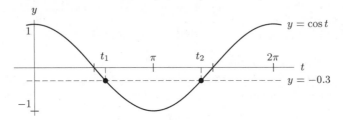

Figure 6.18: The angles t_1 and t_2 are the two solutions to $\cos t = -0.3$ for $0 \le t \le 2\pi$

(b) If your calculator is in radian mode, you should find

$$\cos^{-1}(-0.3) \approx 1.875,$$

which is one of the values we found in part (a) by using a graph. Using the $\boxed{\cos^{-1}}$ key gives only one of the solutions to a trigonometric equation. We find the other solutions by using the symmetry of the unit circle. Figure 6.19 shows that if $t_1 \approx 1.875$ is the first solution, then the second solution is

$$t_2 = 2\pi - t_1$$
$$\approx 2\pi - 1.875 \approx 4.408.$$

Thus, the two solutions are $t \approx 1.88$ and $t \approx 4.41$.

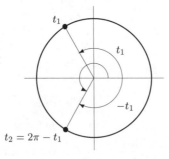

Figure 6.19: By the symmetry of the unit circle, $t_2 = 2\pi - t_1$

5. We divide both sides by 3 to get $\cos \theta = 0.238$. We use the inverse cosine function on a calculator to get $\theta = 1.330$.

9. Using a graphing calculator we graph $y = \sin \theta$ and $y = 0.75$ and look for intersections where $0 \leq \theta \leq \pi$. (See Figure 6.20.) We find $\theta \approx 0.848$ and $\theta \approx 2.294$. The exact values of the solutions are $\theta = \sin^{-1}(3/4)$ and $\theta = \pi - \sin^{-1}(3/4)$.

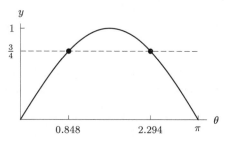

Figure 6.20

13. Since every angle that is a multiple of 2π different from our original angle gives the same point on the unit circle, we first find the angle closest to zero after subtracting as many multiples of 2π as necessary. In this case, since $11\pi/4 = 2\pi + 3\pi/4$, we subtract 2π once, to give us $3\pi/4$. Since this angle is in the second quadrant, we the reference angle is $\pi/4$.

17. Since every angle that is a multiple of 2π different from our original angle gives the same point on the unit circle, we first find the angle closest to zero after adding as many multiples of 2π as necessary. In this case, we add 2π four times, to give us $-22 + 8\pi \approx 3.133$, which is in the second quadrant, between $\pi/2$ and π. Therefore, our reference angle is 0.009.

21. Graph $y = \tan x$ on $0 \leq x \leq 2\pi$ and locate the two points with y-coordinate 2.8. The x-coordinates of these points are approximately $x = 1.228$ and $x = 4.369$. See Figure 6.21.

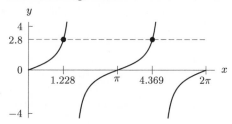

Figure 6.21

Problems

25. One solution is $\theta = \cos^{-1}(\sqrt{3}/2) = \pi/6$, and a second solution is $11\pi/6$, since $\cos(11\pi/6) = \sqrt{3}/2$. All other solutions are found by adding integer multiples of 2π to these two solutions. See Figure 6.22.

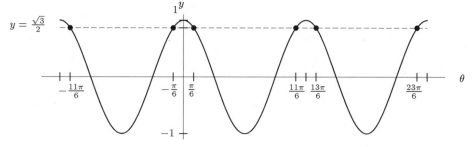

Figure 6.22

29. From Figure 6.23 we can see that the solutions lie on the intervals $\frac{\pi}{8} < t < \frac{\pi}{4}, \frac{3\pi}{4} < t < \frac{7\pi}{8}, \frac{9\pi}{8} < t < \frac{5\pi}{4}$ and $\frac{7\pi}{4} < t < \frac{15\pi}{8}$.
Using the trace mode on a calculator, we can find approximate solutions $t = 0.52$, $t = 2.62$, $t = 3.67$ and $t = 5.76$.

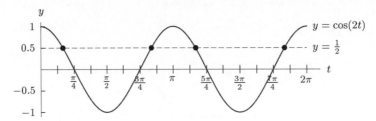

Figure 6.23

For a more precise answer we solve $\cos(2t) = \frac{1}{2}$ algebraically. To find $2t = \arccos(1/2)$. One solution is $2t = \pi/3$. But $2t = 5\pi/3$, $7\pi/3$, and $11\pi/3$ are also angles that have a cosine of $1/2$. Thus $t = \pi/6$, $5\pi/6$, $7\pi/6$, and $11\pi/6$ are the solutions between 0 and 2π.

33. (a) The maximum is \$100,000 and the minimum is \$20,000. Thus in the function

$$f(t) = A\cos(B(t-h)) + k,$$

the midline is

$$k = \frac{100,000 + 20,000}{2} = \$60,000$$

and the amplitude is

$$A = \frac{100,000 - 20,000}{2} = \$40,000.$$

The period of this function is 12 since the sales are seasonal. Since

$$\text{period } = 12 = \frac{2\pi}{B},$$

we have

$$B = \frac{\pi}{6}.$$

The company makes its peak sales in mid-December, which is month -1 or month 11. Since the regular cosine curve hits its peak at $t = 0$ while ours does this at $t = -1$, we find that our curve is shifted horizontally 1 unit to the left. So we have

$$h = -1.$$

So the sales function is

$$f(t) = 40,000\cos\left(\frac{\pi}{6}(t+1)\right) + 60,000 = 40,000\cos\left(\frac{\pi}{6}t + \frac{\pi}{6}\right) + 60,000.$$

(b) Mid-April is month $t = 3$. Substituting this value into our function, we get

$$f(3) = \$40,000.$$

(c) To solve $f(t) = 60,000$ for t, we write

$$60,000 = 40,000\cos\left(\frac{\pi}{6}t + \frac{\pi}{6}\right) + 60,000$$

$$0 = 40,000\cos\left(\frac{\pi}{6}t + \frac{\pi}{6}\right)$$

$$0 = \cos\left(\frac{\pi}{6}t + \frac{\pi}{6}\right).$$

Therefore, $\left(\frac{\pi}{6}t + \frac{\pi}{6}\right)$ equals $\frac{\pi}{2}$ or $\frac{3\pi}{2}$. Solving for t, we get $t = 2$ or $t = 8$. So in mid-March and mid-September the company has sales of \$60,000 (which is the average or midline sales value.)

37. **(a)** $\arccos(0.5) = \pi/3$
   (b) $\arccos(-1) = \pi$
   (c) $\arcsin(0.1) \approx 0.1$

41. **(a)** The domain of f is $-1 \leq x \leq 1$, so $\sin^{-1}(x)$ is defined only if x is between -1 and 1 inclusive. The range is between $-\pi/2$ and $\pi/2$ inclusive.
   (b) The domain of g is $-1 \leq x \leq 1$, so $\cos^{-1}(x)$ is defined only if x is between -1 and 1 inclusive. The range is between 0 and π inclusive.
   (c) The domain of h is all real numbers, so $\tan^{-1}(x)$ is defined for all values of x. The range is between but not including $-\pi/2$ and $\pi/2$.

45. **(a)** The value of $\sin t$ will be between -1 and 1. This means that $k \sin t$ will be between $-k$ and k. Thus, $t^2 = k \sin t$ will be between 0 and k. So

$$-\sqrt{k} \leq t \leq \sqrt{k}.$$

   (b) Plotting $2 \sin t$ and t^2 on a calculator, we see that $t^2 = 2 \sin t$ for $t = 0$ and $t \approx 1.40$.
   (c) Compare the graphs of $k \sin t$, a sine wave, and t^2, a parabola. As k increases, the amplitude of the sine wave increases, and so the sine wave intersects the parabola in more points.
   (d) Plotting $k \sin t$ and t^2 on a calculator for different values of k, we see that if $k \approx 20$, this equation will have a negative solution at $t \approx -4.3$, but that if k is any smaller, there will be no negative solution.

Solutions for Chapter 6 Review

Exercises

1. The function completes one and a half oscillations in 9 units of t, so the period is $9/1.5 = 6$, the amplitude is 5, and the midline is 0.

5. In order to change from degrees to radians, we multiply the number of degrees by $\pi/180$, so we have $330 \cdot \pi/180$, giving $\frac{11}{6}\pi$ radians.

9. In order to change from radians to degrees, we multiply the number of radians by $180/\pi$, so we have $\frac{3}{2}\pi \cdot 180/\pi$, giving 270 degrees.

13. If we go around six times, we make six full circles, which is $2\pi \cdot 6 = 12\pi$ radians. Since we're going in the negative direction, we have -12π radians.

17. The arc length, s, corresponding to an angle of θ radians in a circle of radius r is $s = r\theta$. In order to change from degrees to radians, we multiply the number of degrees by $\pi/180$, so we have $17 \cdot \pi/180$, giving $\frac{17}{180}\pi$ radians. Thus, our arc length is $6.2 \cdot 17\pi/180 \approx 1.840$.

21. The midline is 7. The amplitude is 1. The period is 2π.

25. The amplitude is 1, the period is $2\pi/2 = \pi$, the phase shift is $-\pi/2$, and

$$\text{Horizontal shift} = -\frac{\pi/2}{2} = -\frac{\pi}{4}.$$

Since the horizontal shift is negative, the graph of $y = \cos(2t)$ is shifted $\pi/4$ units to the left to give the graph in Figure 6.24.

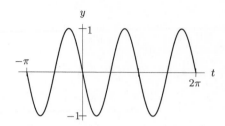

Figure 6.24: $y = \cos(2t + \pi/2)$

29. Since $\tan(-2\pi/3) = \sin(-2\pi/3)/\cos(-2\pi/3)$, we know that $\tan(-2\pi/3) = (-\sqrt{3}/2)/(-1/2) = \sqrt{3}$.

33. From the figure, we see that

$$\sin x = \frac{0.83}{1},$$

so

$$x = \sin^{-1}(0.83).$$

Using a calculator, we find that $x = \sin^{-1}(0.83) \approx 0.979$.

Problems

37. Graph $y = \cos(4\theta)$ and $y = -1$ and find the θ-coordinates of the intersection points. There are four solutions: $\theta \approx \pi/4$, $\theta \approx 3\pi/4$, $\theta \approx 5\pi/4$, and $\theta \approx 7\pi/4$. See Figure 6.25.

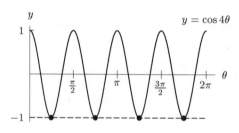

Figure 6.25

41. We first solve for $\tan \alpha$,

$$4 \tan \alpha + 3 = 2$$
$$4 \tan \alpha = -1$$
$$\tan \alpha = -\frac{1}{4}$$
$$\alpha = 2.897, \ 6.038.$$

45. We know $r = 3960$ and $\theta = 1°$. Change θ to radian measure and use $s = r\theta$.

$$s = 3960(1)\left(\frac{\pi}{180}\right) \approx 69.115 \text{ miles.}$$

49. The graph resembles that of a cosine function with a maximum of 7, a minimum of -3 and a period of π, but no horizontal shift. Thus a possible formula is $f(x) = 5\cos(2x) + 2$.

53.

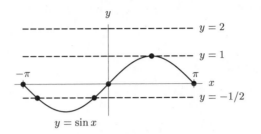

Figure 6.26

Figure 6.26 shows the graph of $y = \sin x$ for $-\pi \le x \le \pi$, together with the horizontal lines $y = 1$, $y = -1/2$, $y = 2$ and $y = 0$ (the x-axis).

(a) Since the line $y = 1$ cuts the graph once, the equation $\sin x = 1$ has one solution for $-\pi \le x \le \pi$.

(b) Since the line $y = -1/2$ cuts the graph twice, the equation $\sin x = -1/2$ has two solutions for $-\pi \le x \le \pi$.

(c) Since the line $y = 0$ (the x-axis) cuts the graph three times, the equation $\sin x = 0$ has three solutions for $-\pi \le x \le \pi$.

(d) Since the line $y = 2$ does not cut the graph at all, the equation $\sin x = 2$ has no solutions for $-\pi \le x \le \pi$. (In fact, this equation has no solutions for any other x-values either.)

57. The data shows the period is about 0.6 sec. The angular frequency is $B = 2\pi/0.6 \approx 10.472$. The amplitude is (high − low)$/2 = (180 - 120)/2 = 60/2 = 30$ cm. The midline is (Minimum) + (Amplitude) $= 120 + 30 = 150$ cm. The weight starts in a low position so it is a quarter cycle behind the sine; that is, the phase shift is $\pi/2$:

$$y = 30\sin\left(105t - \frac{\pi}{2}\right) + 150.$$

CHECK YOUR UNDERSTANDING

1. True, because $\sin x$ is an odd function.

5. True, by the sum of angles identity.

9. False, since $\cos(1/x)$ is undefined at $x = 0$, whereas $(\cos 1)/(\cos x)$ is 1 at $x = 0$.

13. False, since $\sin(\pi x)$ has period $2\pi/\pi = 2$.

17. False. The amplitude is half the difference between its maximum and minimum values.

21. True. The point $(1, 0)$ on the unit circle is the starting point to measure angles so $\theta = 0°$.

25. True. Because P and Q have the same x-coordinates.

29. False. One radian is about 57 degrees.

33. False. Use $s = r\theta$ to get, $s = 3 \cdot (\pi/3) = \pi$.

37. False. $\sin\frac{\pi}{6} = \frac{1}{2}$.

41. True. The period appears to be about 4.

45. False. The period of g is only half as long as the period of f.

49. False. The maximum y-value is when $\cos x = 1$. It is 35.

53. False. The amplitude is positive 2.

57. True. The period is $2\pi/B$.

61. False. Numerically, we could check the equation at $x = 0$ to find $y = -0.5\cos(2 \cdot 0 + \frac{\pi}{3}) + 1 = 0.75$, but the graph at $x = 0$ shows $y = 1$.

65. False. The tangent is undefined at $\frac{\pi}{2}$. Note that $\tan x$ approaches $+\infty$ as x approaches $\pi/2$ from the left and approaches $-\infty$ as x approaches $\pi/2$ from the right.

69. True. The value of $\csc \pi = 1/\sin \pi = 1/0$ is undefined.

73. True. $\sin(\pi/3) = \sqrt{3}/2$.

77. True. If $\cos t = 1$, then $\sin t = 0$ so $\tan t = 0/1 = 0$.

81. False. Any angle $\theta = \frac{\pi}{4} + n\pi$, or $\theta = -\frac{\pi}{4} + n\pi$ with n an integer, will have the same cosine value.

85. False; for example, $\cos\left(\frac{\pi}{4}\right) = \cos\left(\frac{7\pi}{4}\right)$ but $\sin\left(\frac{\pi}{4}\right) = -\sin\left(\frac{7\pi}{4}\right)$.

Solutions to Tools for Chapter 6

1. By the Pythagorean theorem, the hypotenuse has length $\sqrt{1^2 + 2^2} = \sqrt{5}$.

   (a) $\tan \theta = \dfrac{\text{opposite}}{\text{adjacent}} = \dfrac{2}{1} = 2$.

   (b) $\sin \theta = \dfrac{\text{opposite}}{\text{hypotenuse}} = \dfrac{2}{\sqrt{5}}$.

   (c) $\cos \theta = \dfrac{\text{adjacent}}{\text{hypotenuse}} = \dfrac{1}{\sqrt{5}}$.

5. By the Pythagorean Theorem, we know that the third side must be $\sqrt{7^2 - 2^2} = \sqrt{45}$.

   (a) Since $\sin \theta$ is opposite side over hypotenuse, we have $\sin \theta = \sqrt{45}/7$.

   (b) Since $\cos \theta$ is adjacent side over hypotenuse, we have $\cos \theta = 2/7$.

   (c) Since $\tan \theta$ is opposite side over adjacent side, we have $\tan \theta = \sqrt{45}/2$.

9. By the Pythagorean Theorem, we know that the third side must be $\sqrt{11^2 - 2^2} = \sqrt{117}$.

   (a) Since $\sin \theta$ is opposite side over hypotenuse, we have $\sin \theta = \sqrt{117}/11$.

   (b) Since $\cos \theta$ is adjacent side over hypotenuse, we have $\cos \theta = 2/11$.

   (c) Since $\tan \theta$ is opposite side over adjacent side, we have $\tan \theta = \sqrt{117}/2$.

13. Since $\cos 37° = 6/r$, we have $r = 6/\cos 37°$. Similarly, since $\tan 37° = q/6$, we have $q = 6 \tan 37°$.

17. Using $\tan 13° = \dfrac{\text{height}}{200}$ to find the height we get

$$\text{height} = 200 \tan 13° \approx 46.174 \text{feet}.$$

Using $\cos 13° = \dfrac{200}{\text{incline}}$ to find the incline we get

$$\text{incline} = 200/\cos 13° \approx 205.261 \text{feet}.$$

21. Draw a picture as in Figure 6.27. The angle that we want is labeled θ in this picture. We see that $\tan \theta = \dfrac{17.3}{10} = 1.73$. Evaluating $\tan^{-1}(1.73)$ on a calculator, we get $\theta \approx 59.971°$.

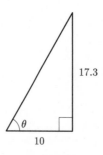

17.3

θ

10

Figure 6.27

25. Let d be the horizontal distance from the airplane to the arch. See Figure 6.28. Then, $\tan\theta = 35000/d$, or $d = 35000/\tan\theta$ feet.

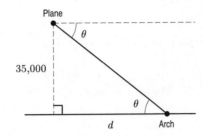

Figure 6.28

29. Since the distance from P to A is $\dfrac{50}{\tan 42°}$ and the distance from P to B is $\dfrac{50}{\tan 35°}$,

$$d = \frac{50}{\tan 35°} - \frac{50}{\tan 42°} \approx 15.877 \text{ feet.}$$

CHAPTER SEVEN

Solutions for Section 7.1

Exercises

1. We begin by using the law of cosines to find side c:

$$c^2 = 7^2 + 6^2 - 2 \cdot 6 \cdot 7 \cos 12°$$
$$c^2 = 2.836$$
$$c = 1.684.$$

We can now use the law of sines to find the other two angles.

$$\frac{\sin B}{6} = \frac{\sin 12°}{1.684}$$
$$\sin B = 6\frac{\sin 12°}{1.684}$$
$$B = \sin^{-1} 0.741$$
$$B = 47.801°.$$

Therefore, $A = 180° - 12° - 47.801 = 120.199°$.

5. We begin by finding the angle C, which is $180° - 13° - 25° = 142°$.
We can now use the law of sines to find the other two sides.

$$\frac{b}{\sin 25°} = \frac{4}{\sin 142°}$$
$$b = \sin 25° \cdot \frac{4}{\sin 142°}$$
$$b = 2.746.$$

Similarly,

$$\frac{a}{\sin 13°} = \frac{4}{\sin 142°}$$
$$a = \sin 13° \cdot \frac{4}{\sin 142°}$$
$$a = 1.462.$$

9. We begin by finding the angle B, which is $180° - 77° - 14° = 89°$.
We can now use the law of sines to find the other two sides.

$$\frac{a}{\sin 14°} = \frac{22}{\sin 77°}$$
$$a = \sin 14° \cdot \frac{22}{\sin 77°}$$
$$a = 5.462.$$

Similarly,

$$\frac{b}{\sin 89°} = \frac{22}{\sin 77°}$$

$$b = \sin 89° \cdot \frac{22}{\sin 77°}$$

$$b = 22.575.$$

13. First, we recognize that it is possible that there are two triangles, since we may have the ambiguous case. However, since 95° is greater than 90°, there are no other obtuse angles possible, so there is but one possible triangle.

We begin by finding the angle C using the law of sines:

$$\frac{\sin C}{10} = \frac{\sin 95°}{5}$$

$$\sin C = 10 \cdot \frac{\sin 95°}{5}$$

$$C = \sin^{-1} 1.992.$$

Since there is no arcsine of 1.992, we notice that there is a problem. There are no solutions. We could have seen this before because the longest side is always across from the largest angle, and since C cannot be greater than 95°, the side across from it (10) cannot be longer than the side across from 95°. Since it is bigger, no triangle fulfills the conditions given.

17. In Figure 7.1, use the Law of Sines:

$$\frac{\sin(30°)}{259} = \frac{\sin \beta}{510}$$

to obtain $\sin \beta \approx 0.9846$ and use $\sin^{-1}$ to find $\beta_1 \approx 79.917$ or $\beta_2 \approx 100.083$. We then know $\alpha_1 = 180° - 30° - 79.917° \approx 70.083°$, or $\alpha_2 = 180° - 30° - 100.083° \approx 49.917°$. We can use the value of α and the Law of Sines to find the length of side a:

$$\frac{a_1}{\sin(70.083°)} = \frac{259}{\sin 30°}, \quad \text{or} \quad \frac{a_2}{\sin(49.917°)} = \frac{259}{\sin 30°}.$$
$$a_1 \approx 487.016 \text{ ft} \qquad\qquad a_2 \approx 396.330 \text{ ft}$$

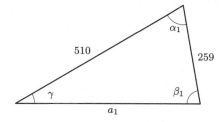 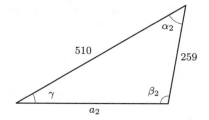

Figure 7.1

Problems

21.

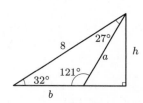

Figure 7.2

(a) $\frac{\sin 121°}{8} = \frac{\sin 32°}{a}$, so $a = \frac{8 \sin 32°}{\sin 121°} \approx 4.946$. Similarly, $b = \frac{8 \sin 27°}{\sin 121°} \approx 4.237$.

(b) Construct an altitude h as in Figure 7.2. We have $\sin 32° = \frac{h}{8}$, so $h = 8 \sin 32° \approx 4.239$. Then area of the triangle is $\frac{1}{2}(4.237)(4.239) = 8.981$.

25. Using the formula for arc length, $s = r\theta$, we have

$$\text{Length of the arc} = 5 \cdot \frac{30\pi}{180} \approx 2.617994 \text{ feet.}$$

Using the Law of Cosines we can solve for the chord length:

$$\text{Length of chord} = \sqrt{5^2 + 5^2 - 2(5)(5)\cos 20°} = \sqrt{50 - 50\cos 30°} \approx 2.588190 \text{ feet.}$$

29. Figure 7.3 shows the triangle; we want to find x. The other two angles are $(180 - 39)°/2 = 70.5°$. Using the Law of Sines

$$\frac{425}{\sin 39°} = \frac{x}{\sin 70.5°}$$

$$x = \frac{425\sin 70.5°}{\sin 39°} = 636.596 \text{ feet.}$$

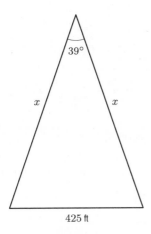

Figure 7.3

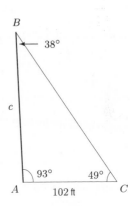

Figure 7.4

33. From Figure 7.4, we see $\angle ABC = 180° - 93° - 49° = 38°$. Using the Law of Sines, we have

$$\frac{102}{\sin 38°} = \frac{c}{\sin 49°}$$

$$c = \frac{102\sin 49°}{\sin 38°} = 125.037 \text{ feet.}$$

37. One way to organize this situation is to use the abbreviations from high school geometry. The six possibilities are { SSS, SAS, SSA, ASA, AAS, AAA }.

SSS Knowing all three sides allows us to find the angles by using the Law of Cosines.

SAS Knowing two sides and the included angle allows us to find the third side length by using the Law of Cosines. We can then use the SSS procedure.

SSA Knowing two sides but not the included angle is called the ambiguous case, because there could be two different solutions. Use the Law of Sines to find one of the missing angles, which, because we use the arcsin, may give two values. Or, use the Law of Cosines, which produces a quadratic equation that may also give two values. Treating these cases separately we can continue to find all sides and angles using the SAS procedure.

ASA Knowing two angles allows us to easily find the third angle. Use the Law of Sines to find each side.

AAS Find the third angle and then use the Law of Sines to find each side.

AAA This has an infinite number of solutions because of similarity of triangles. Once one side is known, then the ASA or AAS procedure can be followed.

Solutions for Section 7.2

Exercises

1. The relevant identities are $\cos^2\theta + \sin^2\theta = 1$ and $\cos 2\theta = \cos^2\theta - \sin^2\theta = 2\cos^2\theta - 1 = 1 - 2\sin^2\theta$. See Table 7.1.

Table 7.1

θ in rad.	$\sin^2\theta$	$\cos^2\theta$	$\sin 2\theta$	$\cos 2\theta$
1	0.708	0.292	0.909	−0.416
$\pi/2$	1	0	0	−1
2	0.827	0.173	−0.757	−0.654
$5\pi/6$	1/4	3/4	$-\sqrt{3}/2$	1/2

5. Writing $\sin 2\alpha = 2\sin\alpha\cos\alpha$, we have

$$\frac{\sin 2\alpha}{\cos\alpha} = \frac{2\sin\alpha\cos\alpha}{\cos\alpha} = 2\sin\alpha.$$

9. Combining terms and using $\cos^2\phi + \sin^2\phi = 1$, we have

$$\frac{\cos\phi - 1}{\sin\phi} + \frac{\sin\phi}{\cos\phi + 1} = \frac{(\cos\phi - 1)(\cos\phi + 1) + \sin^2\phi}{\sin\phi(\cos\phi + 1)} = \frac{\cos^2\phi - 1 + \sin^2\phi}{\sin\phi(\cos\phi + 1)} = \frac{0}{\sin\phi(\cos\phi + 1)} = 0$$

Problems

13. We have

$$\tan 2t = \frac{\sin 2t}{\cos 2t} = \frac{2\sin t\cos t}{\cos^2 t - \sin^2 t}.$$

Dividing both top and bottom by $\cos^2 t$ gives

$$\tan 2t = \frac{\dfrac{2\sin t\cos t}{\cos^2 t}}{\dfrac{\cos^2 t - \sin^2 t}{\cos^2 t}} = \frac{2\tan t}{1 - \tan^2 t}.$$

17. In order to get tan to appear, divide by $\cos x\cos y$:

$$\frac{\sin x\cos y + \cos x\sin y}{\cos x\cos y - \sin x\sin y} = \frac{\dfrac{\sin x\cos y}{\cos x\cos y} + \dfrac{\cos x\sin y}{\cos x\cos y}}{\dfrac{\cos x\cos y}{\cos x\cos y} - \dfrac{\sin x\sin y}{\cos x\cos y}} = \frac{\tan x + \tan y}{1 - \tan x\tan y}$$

21. Not an identity. False for $x = \pi/2$.

25. Identity. $\dfrac{2\tan x}{1 + \tan^2 x} \cdot \dfrac{\cos^2 x}{\cos^2 x} = \dfrac{2\sin x\cos x}{\cos^2 x + \sin^2 x} = \dfrac{\sin 2x}{1} = \sin 2x.$

29. Note the hypotenuse of the triangle is $\sqrt{1 + y^2}$.

(a) $y = \dfrac{y}{1} = \tan\theta.$

(b) $\cos\phi = \sin(\pi/2 - \phi) = \sin\theta.$

(c) Since $\cos\theta = \dfrac{1}{\sqrt{1 + y^2}}$, we have $\sqrt{1 + y^2} = \dfrac{1}{\cos\theta}$, or $1 + y^2 = \left(\dfrac{1}{\cos\theta}\right)^2$. (Alternatively, $1 + y^2 = 1 + \tan^2\theta$.)

(d) Triangle area $= \dfrac{1}{2}(\text{base})(\text{height}) = \dfrac{1}{2}(1)(y)$. But $y = \tan\theta$, so the area is $\frac{1}{2}\tan\theta$.

33. We have $\cos\theta = x/3$, so $\sin\theta = \sqrt{1-(x/3)^2} = \frac{\sqrt{9-x^2}}{3}$. Therefore,

$$\sin 2\theta = 2\sin\theta\cos\theta = 2\left(\frac{\sqrt{9-x^2}}{3}\right)\left(\frac{x}{3}\right) = \frac{2x}{9}\sqrt{9-x^2}.$$

37. First use $\sin(2x) = 2\sin x\cos x$, where $x = 2\theta$. Then

$$\sin(4\theta) = \sin(2x) = 2\sin(2\theta)\cos(2\theta).$$

Since $\sin(2\theta) = 2\sin\theta\cos\theta$ and $\cos(2\theta) = 2\cos^2\theta - 1$, we have

$$\sin 4\theta = 2(2\sin\theta\cos\theta)(2\cos^2\theta - 1).$$

Solutions for Section 7.3

Exercises

1. We have $A = \sqrt{8^2 + 6^2} = \sqrt{100} = 10$. Since $\cos\phi = 8/10 = 0.8$ and $\sin\phi = 6/10 = 0.6$ are both positive, ϕ is in the first quadrant. Thus,

$$\tan\phi = \frac{6}{8} = 0.75 \quad\text{and}\quad \phi = \tan^{-1}(0.75) = 0.644,$$

so $8\sin t + 6\cos t = 10\sin(t + 0.644)$.

5. Write $\sin 15° = \sin(45° - 30°)$, and then apply the appropriate trigonometric identity.

$$\sin 15° = \sin(45° - 30°)$$
$$= \sin 45°\cos 30° - \sin 30°\cos 45°$$
$$= \frac{\sqrt{6}}{4} - \frac{\sqrt{2}}{4}$$

Similarly, $\sin 75° = \sin(45° + 30°)$.

$$\sin 75° = \sin(45° + 30°)$$
$$= \sin 45°\cos 30° + \sin 30°\cos 45°$$
$$= \frac{\sqrt{6}}{4} + \frac{\sqrt{2}}{4}$$

Also, note that $\cos 75° = \sin(90° - 75°) = \sin 15°$, and $\cos 15° = \sin(90° - 15°) = \sin 75°$.

9. Since $285°$ is $240° + 45°$, we can use the sum-of-angle formula for cosine and say that

$$\cos 285 = \cos 240\cos 45 - \sin 240\sin 45 = (-1/2)(\sqrt{2}/2) - (-\sqrt{3}/2)(\sqrt{2}/2) = (-\sqrt{2} + \sqrt{6})/4.$$

Problems

13. (a) $\cos(t - \pi/2) = \cos t\cos\pi/2 + \sin t\sin\pi/2 = \cos t\cdot 0 + \sin t\cdot 1 = \sin t.$
(b) $\sin(t + \pi/2) = \sin t\cos\pi/2 + \sin\pi/2\cos t = \sin t\cdot 0 + 1\cdot\cos t = \cos t.$

17. We can use the identity $\sin u + \sin v = 2\sin((u+v)/2)\cos((u-v)/2)$. If we put $u = 4x$ and $v = x$ then our equation becomes

$$0 = 2\sin\left(\frac{4x+x}{2}\right)\cos\left(\frac{4x-x}{2}\right)$$
$$= 2\sin\left(\frac{5}{2}x\right)\cos\left(\frac{3}{2}x\right)$$

The product on the right-hand side of this equation will be equal to zero precisely when $\sin(5x/2) = 0$ or when $\cos(3x/2) = 0$. We will have $\sin(5x/2) = 0$ when $5x/2 = n\pi$, for n an integer, or in other words for $x = (2\pi/5)n$. This will occur in the stated interval for $x = 2\pi/5$, $x = 4\pi/5$, $x = 6\pi/5$, and $x = 8\pi/5$. We will have $\cos(3x/2) = 0$ when $3x/2 = n\pi + \pi/2$, that is, when $x = (2n+1)\pi/3$. This will occur in the stated interval for $x = \pi/3$, $x = \pi$, and $x = 5\pi/3$. So the given expression is solved by

$$x = \frac{2\pi}{5}, \frac{4\pi}{5}, \frac{6\pi}{5}, \frac{8\pi}{5}, \frac{\pi}{3}, \pi, \text{and}\frac{5\pi}{3}.$$

21. We manipulate the equation for the average rate of change as follows:

$$\frac{\sin(x+h) - \sin x}{h} = \frac{\sin x \cos h + \sin h \cos x - \sin x}{h}$$
$$= \frac{\sin x \cos h - \sin x}{h} + \frac{\sin h \cos x}{h}$$
$$= \sin x\left(\frac{\cos h - 1}{h}\right) + \cos x\left(\frac{\sin h}{h}\right).$$

25. (a) Since $\triangle CAD$ and $\triangle CDB$ are both right triangles, it is easy to calculate the sine and cosine of their angles:

$$\sin\theta = \frac{c_1}{b}$$
$$\cos\theta = \frac{h}{b}$$
$$\sin\phi = \frac{c_2}{a}$$
$$\cos\phi = \frac{h}{a}.$$

(b) We can calculate the areas of the triangles using the formula Area = Base $\cdot$ Height:

$$\text{Area } \triangle CAD = \frac{1}{2}c_1 \cdot h$$
$$= \frac{1}{2}(b\sin\theta)(a\cos\phi),$$
$$\text{Area } \triangle CDB = \frac{1}{2}c_2 \cdot h$$
$$= \frac{1}{2}(a\sin\phi)(b\cos\theta).$$

(c) We find the area of the whole triangle by summing the area of the two constituent triangles:

$$\text{Area } \triangle ABC = \text{Area } \triangle CAD + \text{Area } CDB$$
$$= \frac{1}{2}(b\sin\theta)(a\cos\phi) + \frac{1}{2}(a\sin\phi)(b\cos\theta)$$
$$= \frac{1}{2}ab(\sin\theta\cos\phi + \sin\phi\cos\theta)$$
$$= \frac{1}{2}ab\sin(\theta + \phi)$$
$$= \frac{1}{2}ab\sin C.$$

Solutions for Section 7.4

Exercises

1.

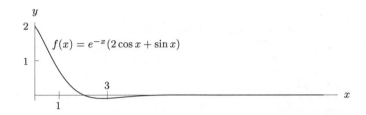

The maximum value of $f(x)$ is 2, which occurs when $x = 0$. The minimum appears to be $y \approx -0.094$, at $x \approx 2.820$.

Problems

5. **(a)** As $x \to \infty$, $\frac{1}{x} \to 0$ and we know $\sin 0 = 0$. Thus, $y = 0$ is the equation of the asymptote.

   (b) As $x \to 0$ and $x > 0$, we have $\frac{1}{x} \to \infty$. This means that for small changes of x the change in $\frac{1}{x}$ is large. Since $\frac{1}{x}$ is a large number of radians, the function will oscillate more and more frequently as x becomes smaller.

   (c) No, because the interval on which $f(x)$ completes a full cycle is not constant as x increases.

   (d) $\sin\left(\frac{1}{x}\right) = 0$ means that $\frac{1}{x} = \sin^{-1}(0) + k\pi$ for k equal to some integer. Therefore, $x = \frac{1}{k\pi}$, and the greatest zero of $f(x) = \sin\frac{1}{x}$ corresponds to the smallest k, that is, $k = 1$. Thus, $z_1 = \frac{1}{\pi}$.

   (e) There are an infinite number of zeros because $z = \frac{1}{k\pi}$ for all $k > 0$ are zeros.

   (f) If $a = \frac{1}{k\pi}$ then the largest zero of $f(x)$ less then a would be $b = \frac{1}{(k+1)\pi}$.

9. **(a)** First consider the height $h = f(t)$ of the hub of the wheel that you are on. This is similar to the basic ferris wheel problem. Therefore $f_1(t) = 25 + 15\sin\left(\frac{\pi}{3}t\right)$, because the vertical shift is 25, the amplitude is 15, and the period is 6. Now the smaller wheel will also add or subtract height depending upon time. The difference in height between your position and the hub of the smaller wheel is given by $f_2(t) = 10\sin\left(\frac{\pi}{2}t\right)$ because the radius is 10 and the period is 4. Finally adding the two together we get:

$$f_1(t) + f_2(t) = f(t) = 25 + 15\sin(\frac{\pi}{3}t) + 10\sin(\frac{\pi}{2}t)$$

   (b)

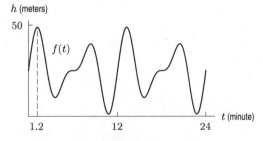

Figure 7.5

Looking at the graph shown in Figure 7.5, we see that $h = f(t)$ is periodic, with period 12. This can be verified by noting

$$f(t + 12) = 25 + 15\sin\left(\frac{\pi}{3}(t + 12)\right) + 10\sin\left(\frac{\pi}{2}(t + 12)\right)$$

$$= 25 + 15 \sin\left(\frac{\pi}{3}t + 4\pi\right) + 10\sin\left(\frac{\pi}{2}t + 6\pi\right)$$

$$= 25 + 15\sin\left(\frac{\pi}{3}t\right) + 10\sin\left(\frac{\pi}{2}t\right)$$

$$= f(t).$$

(c) $h = f(1.2) = 48.776$ m.

Solutions for Section 7.5

Exercises

1. We have $0° < \theta < 90°$. See Figure 7.6.

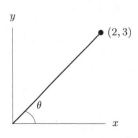

Figure 7.6

5. Quadrant IV.

9. Since $3.2\pi = 2\pi + 1.2\pi$, such a point is in Quadrant III.

13. Since -7 is an angle of -7 radians, corresponding to a rotation of just over 2π, or one full revolution, in the clockwise direction, such a point is in Quadrant IV.

17. With $r = 2$ and $\theta = 5\pi/6$, we find $x = r\cos\theta = 2\cos(5\pi/6) = 2(-\sqrt{3}/2) = -\sqrt{3}$ and $y = r\sin\theta = 2\sin(5\pi/6) = 2(1/2) = 1$.

The rectangular coordinates are $(-\sqrt{3}, 1)$.

21. With $x = -\sqrt{3}$ and $y = 1$, find $r = \sqrt{(-\sqrt{3})^2 + 1^2} = \sqrt{4} = 2$. Find θ from $\tan\theta = y/x = 1/(-\sqrt{3})$. Thus, $\theta = \tan^{-1}(-1/\sqrt{3}) = -\pi/6$. Since $(-\sqrt{3}, 1)$ is in the second quadrant, $\theta = -\pi/6 + \pi = 5\pi/6$. The polar coordinates are $(2, 5\pi/6)$.

Problems

25. Figure 7.7 shows that at 1:30 pm, the polar coordinates of the point H (halfway between 1 and 2 on the clock face) are $r = 3$ and $\theta = 45° = \pi/4$. Thus, the Cartesian coordinates of H are given by

$$x = 3\cos\left(\frac{\pi}{4}\right) = \frac{3\sqrt{2}}{2} \approx 2.121, \quad y = 3\sin\left(\frac{\pi}{4}\right) = \frac{3\sqrt{2}}{2} \approx 2.121.$$

In Cartesian coordinates, $H \approx (2.121, 2.121)$. In polar coordinates, $H = (3, \pi/4)$. In Cartesian coordinates, $M = (0, -4)$. In polar coordinates, $M = (4, 3\pi/2)$.

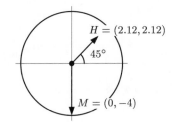

Figure 7.7

29. The region is given by $\sqrt{8} \le r \le \sqrt{18}$ and $\pi/4 \le \theta \le \pi/2$.

33. There will be n loops. See Figures 7.8-7.11.

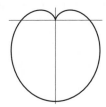

Figure 7.8: $n = 1$

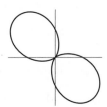

Figure 7.9: $n = 2$

Figure 7.10: $n = 3$

Figure 7.11: $n = 4$

37. Let $0 \le \theta \le 2\pi$ and $3/16 \le r \le 1/2$.

Solutions for Section 7.6

Exercises

1. $2e^{\frac{i\pi}{2}}$

5. $e^{\frac{i3\pi}{2}}$

9. $-5 + 12i$

13. We have $\sqrt{e^{i\pi/3}} = e^{(i\pi/3)/2} = e^{i\pi/6}$, thus $\cos\frac{\pi}{6} + i\sin\frac{\pi}{6} = \frac{\sqrt{3}}{2} + \frac{i}{2}$.

Problems

17. One value of $\sqrt[3]{i}$ is $\sqrt[3]{e^{i\frac{\pi}{2}}} = (e^{i\frac{\pi}{2}})^{\frac{1}{3}} = e^{i\frac{\pi}{6}} = \cos\frac{\pi}{6} + i\sin\frac{\pi}{6} = \frac{\sqrt{3}}{2} + \frac{i}{2}$

21. One value of $(\sqrt{3} + i)^{1/2}$ is
$(2e^{i\frac{\pi}{6}})^{1/2} = \sqrt{2}e^{i\frac{\pi}{12}} = \sqrt{2}\cos\frac{\pi}{12} + i\sqrt{2}\sin\frac{\pi}{12} \approx 1.366 + 0.366i$

25. Substituting $A_2 = i - A_1$ into the second equation gives

$$iA_1 - (i - A_1) = 3,$$

so

$$iA_1 + A_1 = 3 + i$$
$$A_1 = \frac{3+i}{1+i} = \frac{3+i}{1+i} \cdot \frac{1-i}{1-i} = \frac{3 - 3i + i - i^2}{2}$$
$$= 2 - i$$

Therefore $A_2 = i - (2 - i) = -2 + 2i$.

29. Using Euler's formula, we have:

$$e^{i(2\theta)} = \cos 2\theta + i \sin 2\theta$$

On the other hand,

$$e^{i(2\theta)} = \left(e^{i\theta}\right)^2 = (\cos\theta + i\sin\theta)^2 = (\cos^2\theta - \sin^2\theta) + i(2\cos\theta\sin\theta)$$

Equating real parts, we find

$$\cos 2\theta = \cos^2\theta - \sin^2\theta.$$

Solutions for Chapter 7 Review

Exercises

1. (a) First assume $A = 30°$ and $b = 2\sqrt{3}$. Since this is a right triangle, we know that $B = 90° - 30° = 60°$. Now we can determine a by writing

$$\cos A = \frac{\text{adjacent}}{\text{hypotenuse}} = \frac{2\sqrt{3}}{a}.$$

We also know that $\cos A = \frac{\sqrt{3}}{2}$, because A is $30°$. So $\frac{\sqrt{3}}{2} = \frac{2\sqrt{3}}{a}$, which means that $a = 4$. It follows from the Pythagorean theorem that c is $\sqrt{16 - 12} = 2$.

(b) Now assume that $a = 25$ and $c = 24$. The Pythagorean theorem, $a^2 + b^2 = c^2$, implies

$$b = \sqrt{25^2 - 24^2} = 7.$$

To determine angles, we use $\sin A = \frac{\text{opposite}}{\text{hypotenuse}} = \frac{c}{a} = \frac{24}{25}$. So evaluating $\sin^{-1}\left(\frac{24}{25}\right)$ on the calculator, we find that $A = 73.740°$. Therefore, $B = 90° - 73.740° = 16.260°$. (Be sure that your calculator is in degree mode when using the $\sin^{-1}$.)

5. See Figure 7.12. By the Pythagorean theorem, $x = \sqrt{13^2 - 12^2} = 5$.

$$\sin\theta = \frac{12}{13}$$
$$\theta = \sin^{-1}\left(\frac{12}{13}\right)$$
$$\theta \approx 67.380°.$$

Thus $\varphi = 90° - \theta \approx 22.620°$.

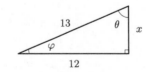

Figure 7.12

9. Writing $\cos 2\phi = 2\cos^2\phi - 1$, we have

$$\frac{\cos 2\phi + 1}{\cos\phi} = \frac{2\cos^2\phi - 1 + 1}{\cos\phi} = \frac{2\cos^2\phi}{\cos\phi} = 2\cos\phi.$$

13. Since $285°$ is $240° + 45°$, we can use the sum-of-angle formula for sine and say that

$$\sin 285 = \sin 240 \cos 45 + \sin 45 \cos 240 = (-\sqrt{3}/2)(\sqrt{2}/2) + (\sqrt{2}/2)(-1/2) = (-\sqrt{6} - \sqrt{2})/4.$$

17. Since $14.4\pi = 7 \cdot 2\pi + 0.4\pi$, such a point is in Quadrant I.

21. With $r = 0$, the point specified is the origin, no matter what the angle measure. So $x = r \cos \theta = 0$ and $y = r \sin \theta = 0$. The rectangular coordinates are $(0, 0)$.

Problems

25. We have $2/7 = \cos 2\theta = 2 \cos^2 \theta - 1$. Solving for $\cos \theta$ gives

$$2 \cos^2 \theta = \frac{9}{7}$$

$$\cos^2 \theta = \frac{9}{14}$$

Since θ is in the first quadrant, $\cos \theta = +\sqrt{9/14} = 3/\sqrt{14}$.

29. See Figure 7.13. They appear to be the same graph. This suggests the truth of the identity $\cos t = \sin(t + \frac{\pi}{2})$.

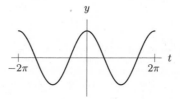

Figure 7.13: Graphs showing $\cos(t) = \sin\left(t + \frac{\pi}{2}\right)$

33. Start with the expression on the left and factor it as the difference of two squares, and then apply the Pythagorean identity to one factor.

$$\begin{aligned}
\sin^4 x - \cos^4 x &= (\sin^2 x)^2 - (\cos^2 x)^2 \\
&= (\sin^2 x - \cos^2 x)(\sin^2 x + \cos^2 x) \\
&= (\sin^2 x - \cos^2 x)(1) \\
&= (\sin^2 x - \cos^2 x).
\end{aligned}$$

37. (a) The side opposite of angle ϕ has length b and the side adjacent to angle ϕ has length a. Therefore,

$$\sin \phi = \frac{\text{side opposite}}{\text{hypotenuse}} = \frac{b}{c}$$

$$\cos \phi = \frac{\text{side adjacent}}{\text{hypotenuse}} = \frac{a}{c}$$

$$\tan \phi = \frac{\text{side opposite}}{\text{side adjacent}} = \frac{b}{a}.$$

(b)

$$\sin \phi = \frac{\text{side opposite } \phi}{\text{hypotenuse}} = \frac{b}{c},$$

$$\cos \theta = \frac{\text{side adjacent to } \theta}{\text{hypotenuse}} = \frac{b}{c}.$$

Thus $\sin \phi = \cos \theta$. Reversing the roles of ϕ and θ one can show $\cos \phi = \sin \theta$ in exactly the same way.

41. Using the Law of Sines on triangle DEF in Figure 7.14, we have

$$\frac{105.2}{\sin 29°} = \frac{ED}{\sin 68°}$$

$$ED = \frac{105.2 \sin 68°}{\sin 29°} = 201.192 \text{ feet.}$$

Total amount of wire needed $= 201.192 + 145.3 + 23.5 + 20 = 389.992$ feet.

Since wire is sold in 100 feet rolls, 4 rolls of wire are needed.

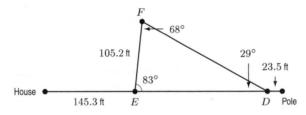

Figure 7.14

CHECK YOUR UNDERSTANDING

1. False. Both acute angles are 45 degrees, and $\sin 45° = \sqrt{2}/2$.

5. True. By the Law of Cosines, we have $p^2 = n^2 + r^2 - 2nr \cos P$, so $\cos P = (n^2 + r^2 - p^2)/(2nr)$.

9. True. Identify the opposite angles as B and L and use the Law of Sines to obtain $\dfrac{LA}{\sin B} = \dfrac{BA}{\sin L}$. Thus $\dfrac{LA}{BA} = \dfrac{\sin B}{\sin L}$.

13. True. Substitute in $\tan \phi = \sin \phi / \cos \phi$ and simplify to see that this is an identity.

17. False. This is not the Pythagorean Identity, since the square is on the variable β. As a counterexample, let $\beta = 1$. Note that $\cos \beta^2 \neq (\cos \beta)^2$ and $\cos^2 \beta = (\cos \beta)^2$.

21. True. There are many ways to prove this identity. We use the identity $\cos 2\theta = \cos^2 \theta - \sin^2 \theta$ to substitute in the right side of the equation. This becomes $\frac{1}{2}(1 - (\cos^2 \theta - \sin^2 \theta))$. Now substitute using $1 - \cos^2 \theta = \sin^2 \theta$ (a form of the Pythagorean identity.) The right side then simplifies to $\sin^2 \theta$ which is the left side.

25. True. Start with the sine sum-of-angle identity:

$$\sin(\theta + \phi) = \sin \theta \cos \phi + \sin \phi \cos \theta$$

and let $\phi = \pi/2$, so

$$\sin(\theta + \pi/2) = \sin \theta \cos(\pi/2) + \sin(\pi/2) \cos \theta.$$

Simplify to

$$\sin(\theta + \pi/2) = \sin \theta \cdot 0 + 1 \cdot \cos \theta = \cos \theta.$$

29. True. We have $f(t) = A \sin(2\pi t + \phi)$ where $A = \sqrt{3^2 + 4^2} = 5$, $\cos \phi = \frac{3}{5}$ and $\sin \phi = \frac{4}{5}$.

33. True. The zeros occur when $\cos(\pi t) + 1 = 0$, so $\cos(\pi t) = -1$. Hence, $\pi t = n\pi$, where n is an odd integer. Thus, $t = n$ is an odd integer.

37. False. Cartesian coordinate values are unique, but polar coordinate values are not. For example, in polar coordinates, $(1, \pi)$ and $(1, 3\pi)$ represent the same point in the xy-plane.

41. True, since $(x - iy)(x + iy) = x^2 + y^2$ is real.

45. False, since $(1 + 2i)^2 = -3 + 4i$.

CHAPTER EIGHT

Solutions for Section 8.1

Exercises

1. To construct a table of values for r, we must evaluate $r(0), r(1), \ldots, r(5)$. Starting with $r(0)$, we have

$$r(0) = p(q(0)).$$

Therefore

$$r(0) = p(5) \qquad \text{(because } q(0) = 5\text{)}$$

Using the table given in the problem, we have

$$r(0) = 4.$$

We can repeat this process for $r(1)$:

$$r(1) = p(q(1)) = p(2) = 5.$$

Similarly,

$$r(2) = p(q(2)) = p(3) = 2$$
$$r(3) = p(q(3)) = p(1) = 0$$
$$r(4) = p(q(4)) = p(4) = 3$$
$$r(5) = p(q(5)) = p(8) = \text{ undefined.}$$

These results have been compiled in Table 8.1.

Table 8.1

x	0	1	2	3	4	5
$r(x)$	4	5	2	0	3	–

5. We want to replace each x in the formula for $f(x)$ with the value of $g(x)$, that is, $\dfrac{1}{x-3}$. The result is $\left(\dfrac{1}{x-3}\right)^2 + 1 = \dfrac{1}{x^2 - 6x + 9} + 1 = \dfrac{1 + x^2 - 6x + 9}{x^2 - 6x + 9} = \dfrac{x^2 - 6x + 10}{x^2 - 6x + 9}$.

9. We take the expression for $g(x)$, namely $\dfrac{1}{x-3}$, and substitute it back into the same expression wherever an x appears. The result is $\dfrac{1}{\frac{1}{x-3} - 3}$. We need to simplify the denominator: $\dfrac{1}{x-3} - 3 = \dfrac{1}{x-3} - \dfrac{3(x-3)}{x-3} = \dfrac{1 - (3x-9)}{x-3} = \dfrac{10 - 3x}{x-3}$.

So, $\dfrac{1}{\frac{1}{x-3} - 3} = \dfrac{1}{\frac{10-3x}{x-3}} = \dfrac{x-3}{10 - 3x}$.

13. $k(n(x)) = (n(x))^2 = \left(\dfrac{2x^2}{x+1}\right)^2 = \dfrac{4x^4}{(x+1)^2}$

Problems

17. The function $f(h(t))$ gives the area of the circle as a function of time, t.

21.

x	-3	-2	-1	0	1	2	3
$f(x)$	0	2	2	0	2	2	0
$g(x)$	0	2	2	0	-2	-2	0
$h(x)$	0	-2	-2	0	-2	-2	0

25. Reading values of the graph, we make an approximate table of values; we use these values to sketch Figure 8.1.

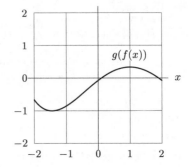

x	-2	-1	0	1	2
$f(x)$	-2	-0.3	0.7	1	0.7
$g(x)$	-0.7	-1	-0.7	0.3	2
$g(f(x))$	-0.7	-0.9	-0.1	0.3	-0.1

Figure 8.1: Graph of $g(f(x))$

29. (a) We have $f(1) = 2$, so $f(f(1)) = f(2) = 4$.
   (b) We have $g(1) = 3$, so $g(g(1)) = g(3) = 1$.
   (c) We have $g(2) = 2$, so $f(g(2)) = f(2) = 4$.
   (d) We have $f(2) = 4$, so $g(f(2)) = g(4) = 0$.

33. The troughs (where the graph is below the x-axis) are reflected about the horizontal axis to become humps. The humps (where the graph is above the x-axis) are unchanged.

37. For $f(x) = \sqrt{x}$

$$\frac{f(x+h) - f(x)}{h} = \frac{\sqrt{x+h} - \sqrt{x}}{h}.$$

41. If $f(x) = u(v(x))$, then one solution is $u(x) = \sqrt{x}$ and $v(x) = 3 - 5x$.

45. One possible solution is $l(x) = u(v(x))$ where $u(x) = 2 + x$ and $v(x) = 1/x$

49. $h(x) = x^3$

53. $j(x) = g(g(x)) = (g(x))^2 + 3 = (x^2 + 3)^2 + 3$.

57. (a) We have

$$f(x) = h\left(g(x)\right) = 3 \cdot 9^x.$$

Since $g(x) = 3^x$, we know that

$$
\begin{aligned}
h\left(g(x)\right) = h(3^x) &= 3 \cdot 9^x \\
&= 3 \cdot (3^2)^x \\
&= 3 \cdot 3^{2x} \\
&= 3(3^x)^2.
\end{aligned}
$$

Since $h(3^x) = 3(3^x)^2$, we know that

$$h(x) = 3x^2.$$

(b) We have

$$f(x) = g\left(j(x)\right) = 3 \cdot 9^x.$$

Since $g(x) = 3^x$, we know that

$$g\left(j(x)\right) = 3^{j(x)} = 3 \cdot 9^x$$
$$= 3 \cdot (3^2)^x$$
$$= 3^1 \cdot 3^{2x}$$
$$= 3^{2x+1}$$

Since $3^{j(x)} = 3^{2x+1}$, we know that

$$j(x) = 2x + 1$$

Solutions for Section 8.2

Exercises

1. It is not invertible.

5. Since $f(x) = (x/4) - (3/2)$ and $g(t) = 4(t + 3/2)$, we have

$$f(g(t)) = \frac{4(t + \frac{3}{2})}{4} - \frac{3}{2} = t + \frac{3}{2} - \frac{3}{2} = t$$
$$g(f(x)) = 4\left(\frac{x}{4} - \frac{3}{2} + \frac{3}{2}\right) = 4\frac{x}{4} = x$$

9. Solve for x in $y = h(x) = 12x^3$:

$$y = 12x^3$$
$$x^3 = \frac{y}{12}$$
$$x = h^{-1}(y) = \sqrt[3]{\frac{y}{12}}.$$

Writing h^{-1} in terms of x gives $h^{-1}(x) = \sqrt[3]{\frac{x}{12}}$.

13. Start with $x = n(n^{-1}(x))$ and substitute $y = n^{-1}(x)$. We have

$$x = n(y)$$
$$x = \log(y - 3)$$
$$10^x = 10^{\log(y-3)}$$
$$10^x = y - 3$$
$$y = 10^x + 3$$

So $y = n^{-1}(x) = 10^x + 3$.

17. Start with $x = f(f^{-1}(x))$ and let $y = f^{-1}(x)$. Then $x = f(y)$ means

$$x = \sqrt{\frac{4 - 7y}{4 - y}}$$
$$x^2 = \frac{4 - 7y}{4 - y}$$
$$x^2(4 - y) = 4 - 7y$$

$$4x^2 - xy^2 = 4 - 7y$$
$$4x^2 - 4 = xy^2 - 7y$$
$$4x^2 - 4 = y(x^2 - 7)$$
$$y = \frac{4x^2 - 4}{x^2 - 7},$$

so $y = f^{-1}(x) = \dfrac{4x^2 - 4}{x^2 - 7}$.

21. Start with $x = q(q^{-1}(x))$ and substitute $y = q^{-1}(x)$. We have

$$x = q(y)$$
$$x = \ln(y + 3) - \ln(y - 5)$$
$$x = \ln \frac{y + 3}{y - 5}$$
$$e^x = e^{\ln\left(\frac{y+3}{y-5}\right)}$$
$$e^x = \frac{y + 3}{y - 5}$$
$$e^x(y - 5) = y + 3$$
$$ye^x - 5e^x = y + 3$$
$$ye^x - y = 3 + 5e^x$$
$$y(e^x - 1) = 3 + 5e^x$$
$$y = \frac{3 + 5e^x}{e^x - 1}$$

So $q^{-1}(x) = \dfrac{3 + 5e^x}{e^x - 1}$.

Problems

25. Solving for t gives

$$P = 10e^{0.02t}$$
$$\frac{P}{10} = e^{0.02t}$$
$$0.02t = \ln\left(\frac{P}{10}\right)$$
$$t = \frac{\ln(P/10)}{0.02} = 50\ln(P/10).$$

The inverse function $t = f^{-1}(P) = 50\ln(P/10)$ gives the time, t, in years at which the population reaches P million.

29. We take logarithms to help solve when x is in the exponent:

$$2^{x+5} = 3$$
$$\ln(2^{x+5}) = \ln 3$$
$$(x + 5)\ln 2 = \ln 3$$
$$x = \frac{\ln 3}{\ln 2} - 5.$$

33. Squaring eliminates square roots:

$$\sqrt{x + \sqrt{x}} = 3$$
$$x + \sqrt{x} = 9$$
$$\sqrt{x} = 9 - x \quad (\text{so} \quad x \le 9)$$
$$x = (9 - x)^2 = 81 - 18x + x^2.$$

So $x^2 - 19x + 81 = 0$. The quadratic formula gives the solutions

$$x = \frac{19 \pm \sqrt{37}}{2}.$$

The only solution is $x = \dfrac{19 - \sqrt{37}}{2}$. The other solution is too large to satisfy the original equation.

37. (a) The graph $y = \sin(t)$ fails the horizontal line test, so we "restrict its domain" to the interval $-\pi/2 \le t \le \pi/2$, resulting in the graph shown in Figure 8.2. This restricted function passes the horizontal line test, and so is invertible. We can define $\sin^{-1}(y)$ to be the inverse of the restricted version of $\sin(t)$.

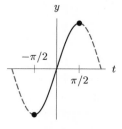

Figure 8.2

(b) The interval $\pi/2 \le t \le 3\pi/2$ can be used because $\sin(t)$ passes the horizontal line test on that interval. Many other answers are possible.

41. If we let $z = \arcsin t$, then we want to simplify $\cos^2 z$. We know that

$$\sin^2 z + \cos^2 z = 1,$$

so we have

$$\cos^2 z = 1 - \sin^2 z.$$

Substituting for z gives

$$\cos^2(\arcsin t) = 1 - \sin^2(\arcsin t).$$

$$\cos^2(\arcsin t) = 1 - (\sin(\arcsin t))^2.$$

Since $\sin(\arcsin t) = t$, we have $\sin^2(\arcsin t) = (\sin(\arcsin t))^2 = t^2$, so

$$\cos^2(\arcsin t) = 1 - t^2.$$

45. (a) We are told that the function is an exponential one, so we know that it must be of the form

$$P(t) = AB^t.$$

Since $P(0) = 150$,

$$P(0) = AB^0 = 150$$
$$= A(1) = 150$$
$$A = 150.$$

Thus

$$P(t) = 150B^t.$$

We know that $P(1) = 165$ so

$$P(1) = 150B^1 = 165$$
$$150B = 165$$
$$B = 1.1.$$

Thus,

$$P(t) = 150(1.1)^t.$$

Checking our answer at $t = 2$ we indeed see that

$$P(2) = AB^2$$
$$= 150(1.1)^2$$
$$= 150(1.21)$$
$$\approx 182.$$

(b) Letting $Y = P(t)$ and solving for t we get

$$Y = 150(1.1)^t$$
$$\frac{Y}{150} = 1.1^t.$$

Taking the log of both sides we get

$$\log\left(\frac{Y}{150}\right) = \log(1.1^t) = t\log(1.1).$$

Dividing both sides by 1.1, we get

$$\frac{\log(\frac{Y}{150})}{\log(1.1)} = t.$$

Recalling that

$$\log\left(\frac{a}{b}\right) = \log(a) - \log(b)$$

we get that

$$t = \frac{\log(Y) - \log(150)}{\log(1.1)}.$$

Since this formula defines the inverse function of $P(t)$ and it is in terms of Y, we can call this function $P^{-1}(Y)$. This function tells us how many years it would take to have Y cows.

$$P^{-1}(Y) = \frac{\log(Y) - \log(150)}{\log(1.1)}.$$

(c) Letting $Y = 400$ in the function $P^{-1}(Y)$ we get

$$P^{-1}(400) = \frac{\log(400) - \log(150)}{\log(1.1)}$$
$$\approx 10.3.$$

Thus it would take roughly 10.3 years for the population of the cattle herd to reach 400. To check that this is indeed the correct answer we can let $t = 10.3$ in our original function $P(t)$

$$P(10.3) = 150(1.1)^{10.3}$$
$$\approx 400.$$

49. (a) Since the amount of alcohol goes from 99% to 98%, you might expect that 1% of the alcohol should be removed. There are originally 99 ml of alcohol, and 1% of 99 = 0.99, and so this would mean that 0.99 ml needs to be removed. (As we shall see, this turns out to be completely wrong!)

(b) If $y = C(x)$, then y is the concentration of alcohol after x ml are removed. Since we want to remove an amount x of alcohol to yield a 98% solution, we have

$$C(x) = 0.98.$$

This means that

$$x = C^{-1}(0.98).$$

(c) We have

$$C^{-1}(0.98) = \frac{99 - 100(0.98)}{1 - 0.98}$$
$$= \frac{99 - 98}{0.02}$$
$$= 50.$$

Thus, we would need to remove 50 ml of alcohol from the 100 ml solution to obtain a 98% solution. This is much more than the 0.99 ml you might have expected. We can double check our answer to be sure. If we begin with 100 ml of solution containing 99 ml of alcohol, and then remove 50 ml of alcohol, we obtain a 50 ml solution containing 49 ml of alcohol. This gives a concentration of $\frac{49}{50} = 98\%$, which is what we wanted.

Solutions for Section 8.3

Exercises

1. To find $f(x)$, we add $m(x)$ and $n(x)$ and simplify: $m(x) + n(x) = 3x^2 - x + 2x = 3x^2 + x = f(x)$.

5. To find $j(x)$, we divide $m(x)$ by $n(x)$ and simplify: $(m(x))/n(x) = (3x^2 - x)/(2x) = 3x/2 - 1/2 = j(x)$.

9. (a) We have $f(x) + g(x) = x + 5 + x - 5 = 2x$.
(b) We have $f(x) - g(x) = x + 5 - (x - 5) = 10$.
(c) We have $f(x)g(x) = (x + 5)(x - 5) = x^2 - 25$.
(d) We have $f(x)/g(x) = (x + 5)/(x - 5)$.

Problems

13. (a) Graphs of f and g are in Figure 8.3.

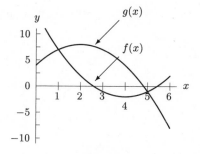

Figure 8.3

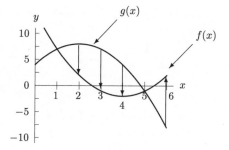

Figure 8.4

(b) Values of $f(x)$, $g(x)$, and $f(x) - g(x)$ are in Table 8.2.

Table 8.2

x	0	1	2	3	4	5	6
$f(x)$	14	7	2	-1	-2	-1	2
$g(x)$	4	7	8	7	4	-1	-8
$f(x) - g(x)$	10	0	-6	-8	-6	0	10

(c) See part (b).

(d) See Figure 8.4.

(e) See Figure 8.5.

(f) $f(x) = x^2 - 8x + 14$, $g(x) = -x^2 + 4x + 4$,
$$f(x) - g(x) = 2x^2 - 12x + 10.$$

(g) See Figure 8.5, where the graph of $f(x) - g(x)$ passes through the points plotted in part (e).

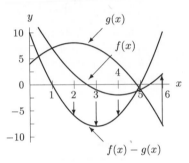

Figure 8.5

17. (a) A formula for $h(x)$ would be
$$h(x) = f(x) + g(x).$$

To evaluate $h(x)$ for $x = 3$, we use this equation:
$$h(3) = f(3) + g(3).$$

Since $f(x) = x + 1$, we know that
$$f(3) = 3 + 1 = 4.$$

Likewise, since $g(x) = x^2 - 1$, we know that
$$g(3) = 3^2 - 1 = 9 - 1 = 8.$$

Thus, we have
$$h(3) = 4 + 8 = 12.$$

To find a formula for $h(x)$ in terms of x, we substitute our formulas for $f(x)$ and $g(x)$ into the equation $h(x) = f(x) + g(x)$:

$$h(x) = \underbrace{f(x)}_{x+1} + \underbrace{g(x)}_{x^2-1}$$
$$h(x) = x + 1 + x^2 - 1 = x^2 + x.$$

To check this formula, we use it to evaluate $h(3)$, and see if it gives $h(3) = 12$, which is what we got before. The formula is $h(x) = x^2 + x$, so it gives
$$h(3) = 3^2 + 3 = 9 + 3 = 12.$$

This is the result that we expected.

(b) A formula for $j(x)$ would be

$$j(x) = g(x) - 2f(x).$$

To evaluate $j(x)$ for $x = 3$, we use this equation:

$$j(3) = g(3) - 2f(3).$$

We already know that $g(3) = 8$ and $f(3) = 4$. Thus,

$$j(3) = 8 - 2 \cdot 4 = 8 - 8 = 0.$$

To find a formula for $j(x)$ in terms of x, we again use the formulas for $f(x)$ and $g(x)$:

$$j(x) = \underbrace{g(x)}_{x^2 - 1} - 2\underbrace{f(x)}_{x + 1}$$
$$= (x^2 - 1) - 2(x + 1)$$
$$= x^2 - 1 - 2x - 2$$
$$= x^2 - 2x - 3.$$

We check this formula using the fact that we already know $j(3) = 0$. Since we have $j(x) = x^2 - 2x - 3$,

$$j(3) = 3^2 - 2 \cdot 3 - 3 = 9 - 6 - 3 = 0.$$

This is the result that we expected.

(c) A formula for $k(x)$ would be

$$k(x) = f(x)g(x).$$

Evaluating $k(3)$, we have

$$k(3) = f(3)g(3) = 4 \cdot 8 = 32.$$

A formula in terms of x for $k(x)$ would be

$$k(x) = \underbrace{f(x)}_{x + 1} \cdot \underbrace{g(x)}_{x^2 - 1}$$
$$= (x + 1)(x^2 - 1)$$
$$= x^3 - x + x^2 - 1$$
$$= x^3 + x^2 - x - 1.$$

To check this formula,

$$k(3) = 3^3 + 3^2 - 3 - 1 = 27 + 9 - 3 - 1 = 32,$$

which agrees with what we already knew.

(d) A formula for $m(x)$ would be

$$m(x) = \frac{g(x)}{f(x)}.$$

Using this formula, we have

$$m(3) = \frac{g(3)}{f(3)} = \frac{8}{4} = 2.$$

To find a formula for $m(x)$ in terms of x, we write

$$m(x) = \frac{g(x)}{f(x)} = \frac{x^2 - 1}{x + 1}$$
$$= \frac{(x + 1)(x - 1)}{(x + 1)}$$
$$= x - 1 \text{ for } x \neq -1$$

We were able to simplify this formula by first factoring the numerator of the fraction $\dfrac{x^2 - 1}{x + 1}$. To check this formula,

$$m(3) = 3 - 1 = 2,$$

which is what we were expecting.

(e) We have

$$n(x) = (f(x))^2 - g(x).$$

This means that

$$
\begin{aligned}
n(3) &= (f(3))^2 - g(3) \\
&= (4)^2 - 8 \\
&= 16 - 8 \\
&= 8.
\end{aligned}
$$

A formula for $n(x)$ in terms of x would be

$$
\begin{aligned}
n(x) &= (f(x))^2 - g(x) \\
&= (x + 1)^2 - (x^2 - 1) \\
&= x^2 + 2x + 1 - x^2 + 1 \\
&= 2x + 2.
\end{aligned}
$$

To check this formula,

$$n(3) = 2 \cdot 3 + 2 = 8,$$

which is what we were expecting.

21. We can find the revenue function as a product:

$$\text{Revenue} = (\text{\# of customers}) \cdot (\text{price per customer}).$$

At the current price, 50,000 people attend every day. Since 2500 customers will be lost for each \$1 increase in price, the function $n(i)$ giving the number of customers who will attend given i one-dollar price increases, is given by $n(i) = 50{,}000 - 2500i$. The price function $p(i)$ giving the price after i one-dollar price increases is given by $p(i) = 15 + i$. The revenue function $r(i)$ is given by

$$
\begin{aligned}
r(i) &= n(i)p(i) \\
&= (50{,}000 - 2500i)(15 + i) \\
&= -2500i^2 + 12{,}500i + 750{,}000 \\
&= -2500(i - 20)(i + 15).
\end{aligned}
$$

The graph $r(i)$ is a downward-facing parabola with zeros at $i = -15$ and $i = 20$, so the maximum revenue occurs at $i = 2.5$ which is halfway between the zeros. Thus, to maximize profits the ideal price is $\$15 + 2.5(\$1.00) = \$15 + \$2.50 = \$17.50$.

25. The graphs are found below.

(a)
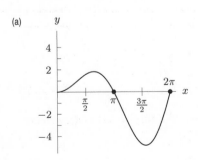

(b)
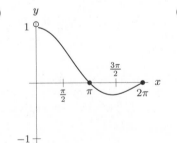

(c)
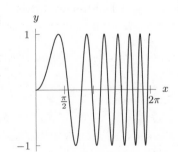

(d)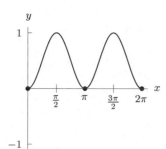

(e)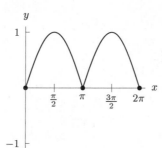

(f)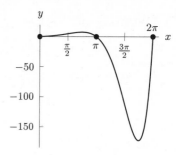

29. The statement is false. For example, if $f(x) = x$ and $g(x) = x^2$, then $f(x) \cdot g(x) = x^3$. In this case, $f(x) \cdot g(x)$ is an odd function, but $g(x)$ is an even function.

33. (a) The function $h_{CA}(t)$ gives the total number of pounds (in 1000s) of strawberries produced in California in year t. See Table 8.3.

Table 8.3 *Values for $h_{CA}(t) = f_{CA}(t) \cdot g_{CA}(t)$*

t	0	1	2	3	4
$h_{CA}(t)$	1,333,400	1,355,200	1,535,100	1,628,400	1,386,000

(b) A formula for $p(t)$ is given by

$$p(t) = \frac{\text{Strawberries in CA and FL}}{\text{Strawberries in US total}}$$
$$= \frac{\text{CA area} \times \text{CA yield} + \text{FL area} \times \text{FL yield}}{\text{US area} \times \text{US yield}}$$
$$= \frac{f_{CA}(t) \cdot g_{CA}(t) + f_{FL}(t) \cdot g_{FL}(t)}{f_{US}(t) \cdot g_{US}(t)}.$$

See Table 8.4.

Table 8.4 *Values for $p(t)$, the fraction of all US strawberries (by weight) grown in Florida and California in year t.*

t	0	1	2	3	4
$p(t)$	0.927	0.908	0.907	0.924	0.892

Solutions for Chapter 8 Review

Exercises

1. Since $f(x) = 3x^2$, we substitute $3x^2$ for x in $g(x)$, giving us $g(f(x)) = 9(3x^2) - 2$, which simplifies to $g(f(x)) = 27x^2 - 2$.

5. Since $g(x) = 9x - 2$, we substitute $9x - 2$ for x in $m(x)$, giving us $m(g(x)) = 4(9x - 2)$, which simplifies to $m(g(x)) = 36x - 8$, which we then substitute for x in $f(x)$, giving $f(m(g(x))) = 3(36x - 8)^2$, which simplifies to $f(m(g(x))) = 3888x^2 - 1728x + 192$.

9. Start with $x = j(j^{-1}(x))$ and substitute $y = j^{-1}(x)$. We have

$$x = j(y)$$
$$x = \sqrt{1 + \sqrt{y}}$$
$$x^2 = 1 + \sqrt{y}$$
$$x^2 - 1 = \sqrt{y}$$
$$(x^2 - 1)^2 = y$$

Therefore,

$$j^{-1}(x) = (x^2 - 1)^2.$$

13. Start with $x = h(h^{-1}(x))$ and substitute $y = h^{-1}(x)$. We have

$$x = h(y)$$
$$x = \log \frac{y + 5}{y - 4}$$
$$10^x = \frac{y + 5}{y - 4}$$
$$10^x(y - 4) = y + 5$$
$$10^x y - 4 \cdot 10^x = y + 5$$
$$10^x y - y = 5 + 4 \cdot 10^x$$
$$y(10^x - 1) = 5 + 4 \cdot 10^x$$
$$y = \frac{5 + 4 \cdot 10^x}{10^x - 1}$$
$$h^{-1}(x) = \frac{5 + 4 \cdot 10^x}{10^x - 1}.$$

Problems

17. Using $f(x) = \dfrac{1}{x + 1}$, we obtain

$$f\left(\frac{1}{x}\right) + \frac{1}{f(x)} = \frac{1}{\frac{1}{x} + 1} + \frac{1}{\frac{1}{x+1}}$$
$$= \frac{1}{\frac{1+x}{x}} + x + 1$$
$$= \frac{x}{1 + x} + x + 1$$

21. $g(x) = \dfrac{1}{x^2}$ and $h(x) = x + 4$

25. Since $h(g(x)) = h(2x + 5)$, we have

$$h(2x + 5) = \frac{2x + 6}{1 + \sqrt{2x + 4}}$$
$$= \frac{2x + 5 + 1}{1 + \sqrt{2x + 5 - 1}}$$

which means

$$h(x) = \frac{x + 1}{1 + \sqrt{x - 1}}$$

29. See Figure 8.6.

$y = g(f(x - 2))$

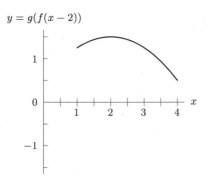

Figure 8.6

$y = f(x) - g(x)$

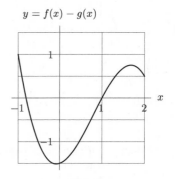

Figure 8.7

33. The graph is in Figure 8.7.

37. Solving for r gives

$$r = \log\left(\frac{I}{I_0}\right)$$

$$r = \log\left(\frac{I}{I_0}\right)$$

$$\frac{I}{I_0} = 10^r$$

$$I = I_0 10^r.$$

The inverse function $f^{-1}(r) = I_0 10^r$ gives the intensity of an earthquake with Richter rating r.

41. (a) Since $P(t)$ is exponential, we know that there are a few ways we could write its formula. One possibility is $P(t) = P_0 b^t$, where b is the annual growth factor. Another is $P(t) = P_0 e^{kt}$ where k is the continuous annual growth rate. If we choose the first, we would have

$$P(t) = P_0 b^t.$$

Since the town triples in size every 7 years, we also know that

$$P(7) = 3P_0.$$

By our formula, we have

$$P(7) = P_0 b^7$$
$$3P_0 = P_0 b^7$$
$$b^7 = 3$$
$$b = 3^{\frac{1}{7}}.$$

Thus, $P(t) = P_0(3^{\frac{1}{7}})^t = P_0(3)^{\frac{t}{7}}$. Since $3^{\frac{1}{7}} \approx 1.170$, we could write $P(t) = P_0(1.170)^t$. If we had used the form $P(t) = P_0 e^{kt}$, we would have found that $P(t) = P_0 e^{0.157t}$.

(b) Since $P(t) = P_0(3^{\frac{1}{7}})^t \approx P_0(1.170)^t$, we know that the town's population increases by about 17% every year.

(c) Letting $x = P(t)$, we solve for t:

$$x = P_0(3)^{t/7}$$
$$3^{t/7} = \frac{x}{P_0}$$

$$\log 3^{t/7} = \log \frac{x}{P_0}$$

$$\left(\frac{t}{7}\right)\log 3 = \log \frac{x}{P_0}$$

$$\frac{t}{7} = \frac{\log(x/P_0)}{\log 3}$$

$$t = \frac{7\log(x/P_0)}{\log 3}.$$

So,

$$P^{-1}(x) = \frac{7\log \frac{x}{P_0}}{\log 3} = \frac{7(\log x - \log P_0)}{\log 3}.$$

$P^{-1}(x)$ is the number of years required for the population to reach x people. Note that there are many different ways to express this formula.

(d) We want to solve $P(t) = 2P_0$. This is equivalent to evaluating $P^{-1}(2P_0)$.

$$P^{-1}(2P_0) = \frac{7\log \frac{2P_0}{P_0}}{\log 3} = \frac{7\log 2}{\log 3} \approx 4.417 \text{ years.}$$

45. Since $2x$ represents twice as much office space as x, the cost of building twice as much space is $f(2x)$. The cost of building x amount of space is $f(x)$, so twice this cost is $2f(x)$. Thus, the contractors statement is expressed

$$f(2x) < 2f(x).$$

49. The inequality $h(f(x)) < x$ tells us that Space can build fewer than x square feet of office space with the money Ace needs to build x square feet. You get more for your money with Ace.

53. This is an increasing function, because if $g(x)$ is a decreasing function, then $-g(x)$ will be an increasing function. Since $f(x) - g(x) = f(x) + [-g(x)]$, $f(x) - g(x)$ can be written as the sum of two increasing functions, and is thus increasing.

CHECK YOUR UNDERSTANDING

1. False. If $x < -3$ then $g(x)$ is not defined, thus $f(g(x))$ is not defined.

5. True. First, $f(1) = 1^2 + 2 = 3$ then $f(3) = 3^2 + 2 = 11$.

9. False. $f(g(x)) = f(\sin x) = (\sin x)^2 = \sin^2 x$.

13. False. None of them has an inverse.

17. True. The graph passes the vertical line test.

21. False, since $\dfrac{h(x)}{f(x)} = \dfrac{x-5}{1/x} = x(x-5)$.

25. True. Since the two sides of the equation are defined only when $x > 0$, we have $f(x)g(x) = \dfrac{1}{x} \cdot \sqrt{x} = \dfrac{1}{\sqrt{x}}$, and

$$f(g(x)) = \frac{1}{\sqrt{x}}.$$

29. False. $4h(2) = 4(2-5) = -12$, but $h(8) = 8 - 5 = 3$.

CHAPTER NINE

Solutions for Section 9.1

Exercises

1. While $y = 6x^3$ is a power function, when we add two to it, it can no longer be written in the form $y = kx^p$, so this is not a power function.

5. Expanding the right side, we get $y = 4x^2 - 16 + 16 = 4x^2$. Thus, this is a power function, $y = 4x^2$.

9. Since the graph is steeper near the origin and less steep away from the origin, the power function is fractional.

13. We use the form $y = kx^p$ and solve for k and p. Using the point $(1, 2)$, we have $2 = k1^p$. Since 1^p is 1 for any p, we know that $k = 2$. Using our other point, we see that

$$17 = 2 \cdot 6^p$$
$$\frac{17}{2} = 6^p$$
$$\ln\left(\frac{17}{2}\right) = p \ln 6$$
$$\frac{\ln(17/2)}{\ln 6} = p$$
$$1.194 \approx p.$$

So $y = 2x^{1.194}$.

17. Substituting into the general formula $c = k/d^2$, we have $45 = k/3^2$ or $k = 405$. So the formula for c is

$$c = \frac{405}{d^2}.$$

When $d = 5$, we get $c = 405/5^2 = 16.2$.

21. Solve for $h(x)$ by taking the ratio of, say, $h(4)$ to $h(\frac{1}{4})$:

$$\frac{h(4)}{h(\frac{1}{4})} = \frac{-1/8}{-32} = \frac{-1}{8} \cdot \frac{-1}{32} = \frac{1}{256}.$$

We know $h(4) = k \cdot 4^p$ and $h(\frac{1}{4}) = k \cdot (\frac{1}{4})^p$. Thus,

$$\frac{h(4)}{h(\frac{1}{4})} = \frac{k \cdot 4^p}{k \cdot (\frac{1}{4})^p} = \frac{4^p}{(\frac{1}{4})^p} = 16^p = \frac{1}{256}.$$

Since $16^p = \frac{1}{256} = \frac{1}{16^2} = 16^{-2}$, $p = -2$. To solve for k, note that $h(4) = k \cdot 4^p = k \cdot 4^{-2} = \frac{k}{16}$. Since $h(4) = -\frac{1}{8}$, we have $\frac{k}{16} = -\frac{1}{8}$. Thus, $k = -2$, which gives $h(x) = -2x^{-2}$.

Problems

25. Notice that,

$$f(-x) = k(-x)^p = k(-1 \cdot x)^p = kx^p \cdot (-1)^p.$$

Now, if p is even, then $(-1)^p = 1$, and if p is odd, then $(-1)^p = -1$. Thus, if p is even,

$$f(-x) = kx^p(-1)^p = kx^p = f(x),$$

and so f is even. But if p is odd, then

$$f(-x) = kx^p(-1)^p = -kx^p = -f(x),$$

and so f is odd.

29. $c(t) = \frac{1}{t}$ is indeed one possible formula. It is not, however, the only one. Because the vertical and horizontal axes are asymptotes for this function, we know that the power p is a negative number and

$$c(t) = kt^p.$$

If $p = -3$ then $c(t) = kt^{-3}$. Since $(2, \frac{1}{2})$ lies on the curve, $\frac{1}{2} = k(2)^{-3}$ or $k = 4$. So, $c(t) = 4t^{-3}$ could describe this function. Similarly, so could $c(t) = 16x^{-5}$ or $c(t) = 64x^{-7}$...

33. The map distance is directly proportional to the actual distance (mileage), because as the actual distance, x, increases, the map distance, d, also increases.

Substituting the values given into the general formula $d = kx$, we have $0.5 = k(5)$, so $k = 0.1$, and the formula is

$$d = 0.1x.$$

When $d = 3.25$, we have $3.25 = 0.1(x)$ so $x = 32.5$. Therefore, towns which are separated by 3.25 inches on the map are 32.5 miles apart.

37. (a) The radius, r, is 0 when $t = 0$ and increases by 200 meters per hour. Thus, after t hours, the radius in meters is given by

$$r = 200t.$$

(b) Since the spill is circular, its area, A, in square meters, is given by

$$A = \pi r^2.$$

Substituting for r from part (a), we have

$$A = \pi(r)^2 = \pi(200t)^2 = 40{,}000\pi t^2.$$

(c) When $t = 7$, the area (in square meters) is

$$A = 40{,}000 \cdot \pi \cdot 7^2 \approx 6{,}157{,}521.601.$$

41. (a) If p is negative, the domain is all x except $x = 0$.
There are no domain restrictions if p is positive.
(b) If p is even and positive, the range is $y \geq 0$.
If p is even and negative, the range is $y > 0$.
If p is odd and positive, the range is all real numbers.
If p is odd and negative, the range is all real numbers except $y = 0$.
(c) If p is even, the graph is symmetric about the y-axis.
If p is odd, the graph is symmetric about the origin.

Solutions for Section 9.2

Exercises

1. This is a polynomial of degree two.

5. This is not a polynomial because $2e^x$ is not a power function.

9. Since the leading term of the polynomial is $16x^3$, the value of y goes to infinity as $x \to \infty$. The graph resembles $y = x^3$.

Problems

13. (a)

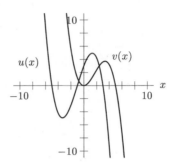

Figure 9.1

The graphs of u and v have the same end behavior. As $x \to -\infty$, both $u(x)$ and $v(x) \to \infty$, and as $x \to \infty$, both $u(x)$ and $v(x) \to -\infty$.

The graphs have different y-intercepts, and u has three distinct zeros. The function v has a repeated zero at $x = 0$.

(b) On the window $-20 \le x \le 20$ by $-1600 \le y \le 1600$, the peaks and valleys of both functions are not distinguishable. Near the origin, the behavior of both functions looks the same. The functions are still distinguishable from one another on the ends.

On the window $-50 \le x \le 50$ by $-25,000 \le y \le 25,000$, the functions are still slightly distinct from one another on the ends—but barely.

On the last window the graphs of both functions appear identical. Both functions look like the function $y = -\frac{1}{5}x^3$.

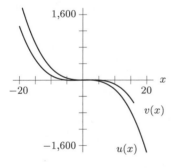
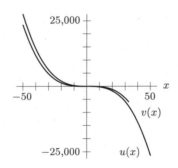
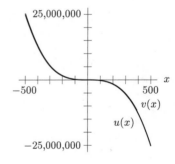

Figure 9.2

17. (a) Using a computer or a graphing calculator, we can get a picture of $f(x)$ like the one in Figure 9.3. On this window f appears to be invertible because it passes the horizontal line test.

(b) Substituting gives

$$f(0.5) = (0.5)^3 + 0.5 + 1 = 1.625.$$

To find $f^{-1}(0.5)$, we solve $f(x) = 0.5$. With a computer or graphing calculator, we trace along the graph of f in Figure 9.4 to find

$$f^{-1}(0.5) \approx -0.424.$$

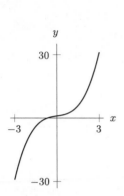

Figure 9.3: $f(x) = x^3 + x + 1$

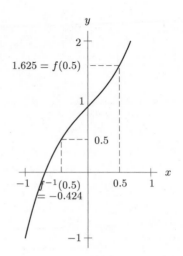

Figure 9.4

21. (a) The graph of the function on the suggested window is shown in Figure 9.5. At $x = 0$ (when Smallsville was founded), the population was 5 hundred people.

(b) The x-intercept for $x > 0$ will show when the population was zero. This occurs at $x \approx 8.44$. Thus, Smallsville became a ghost town in May of 1908.

(c) There are two peaks on the graph on $0 \leq x \leq 10$, but the first occurs before $x = 5$ (i.e.,before 1905). The second peak occurs at $x \approx 7.18$. The population at that point is ≈ 7.9 hundred. So the maximum population was ≈ 790 in February of 1907.

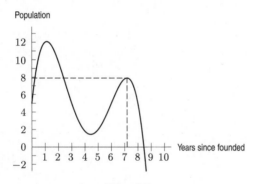

Figure 9.5

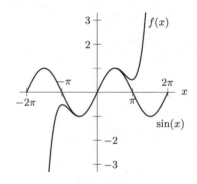

Figure 9.6

25. (a) See Figure 9.6.

(b) The graphs are very similar on the interval

$$\frac{-\pi}{2} \leq x \leq \frac{\pi}{2},$$

and even slightly larger intervals show close similarity.

(c)

$$\sin\left(\frac{\pi}{8}\right) = 0.382683432\cdots$$

$$f\left(\frac{\pi}{8}\right) = \frac{\pi}{8} - \frac{1}{6}\left(\frac{\pi}{8}\right)^3 + \frac{1}{120}\left(\frac{\pi}{8}\right)^5$$
$$= 0.382683717\cdots.$$

As you can see, $f\left(\frac{\pi}{8}\right)$ differs from $\sin\frac{\pi}{8}$ only in the 7th decimal place—that is, by less than 0.0001%.

(d) Since $\sin x$ is periodic with period 2π, we know that $\sin 18 = \sin(18 - 2\pi) = \sin(18 - 4\pi) = \sin(18 - 6\pi) = \cdots$. Notice that $18 - 6\pi = -0.8495\cdots$ is within the interval $-\frac{\pi}{2} \leq x \leq \frac{\pi}{2}$ on which f resembles $\sin x$. Thus, $f(18 - 6\pi) \approx \sin(18 - 6\pi)$, and since $\sin(18 - 6\pi) = \sin 18$, then

$$f(18 - 6\pi) \approx \sin 18.$$

Using a calculator, we find $f(18 - 6\pi) = -0.7510\cdots$, and $\sin(18) = -0.75098\cdots$. Thus, $f(18 - 6\pi)$ is an excellent approximation for $\sin 18$. (In fact, your calculator evaluates trigonometric functions internally using a method similar to the one presented in this problem.)

Solutions for Section 9.3

Exercises

1. The graph shows that $g(x)$ has zeros at $x = -2$, $x = 0$, $x = 2$, $x = 4$. Thus, $g(x)$ has factors of $(x + 2)$, x, $(x - 2)$, and $(x - 4)$, so

$$g(x) = k(x + 2)x(x - 2)(x - 4).$$

Since $g(x) = x^4 - 4x^3 - 4x^2 + 16x$, we see that $k = 1$, so

$$g(x) = x(x + 2)(x - 2)(x - 4).$$

5. Zeros occur where $y = 0$, at $x = -3$, $x = 2$, and $x = -7$.

9. Factoring f gives $f(x) = -5(x + 2)(x - 2)(5 - x)(5 + x)$, so the x intercepts are at $x = -2, 2, 5, -5$.
   The y intercept is at: $y = f(0) = -5(2)(-2)(5)(5) = 500$.

   The polynomial is of fourth degree with the highest powered term $5x^4$. Thus, both ends point upward. A graph of $y = f(x)$ is shown in Figure 9.7.

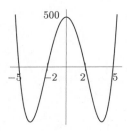

Figure 9.7

Problems

13. The graph appears to have x intercepts at $x = -\frac{1}{2}, 3, 4$, so let

$$y = k(x + \frac{1}{2})(x - 3)(x - 4).$$

The y-intercept is at $(0, 3)$, so substituting $x = 0$, $y = 3$:

$$3 = k(\frac{1}{2})(-3)(-4),$$

which gives $\quad 3 = 6k,$

or $\quad k = \frac{1}{2}.$

Therefore, $y = \frac{1}{2}(x + \frac{1}{2})(x - 3)(x - 4)$ is a possible formula for f.

17. The shape of the graph suggests an odd degree polynomial with a positive leading term. Since the graph crosses the x-axis at -2, there is a factor of $(x+2)$, and since it touches the x-axis at 3, there should be a factor of $(x-3)^2$, giving $y = a(x+2)(x-3)^2$. Since the graph crosses the y-axis at 126, we can find a by substituting $x = 0$:

$$126 = a(0+2)(0-3)^2$$
$$126 = 18a$$
$$7 = a.$$

Thus, a possible polynomial is $y = 7(x+2)(x-3)^2$.

21. We know that $g(-2) = 0$, $g(-1) = -3$, $g(2) = 0$, and $g(3) = 0$. We also know that $x = -2$ is a repeated zero. Thus, let

$$g(x) = k(x+2)^2(x-2)(x-3).$$

Then, using $g(-1) = -3$, gives

$$g(-1) = k(-1+2)^2(-1-2)(-1-3) = k(1)^2(-3)(-4) = 12k,$$

so $12k = -3$, and $k = -\frac{1}{4}$. Thus,

$$g(x) = -\frac{1}{4}(x+2)^2(x-2)(x-3)$$

is a possible formula for g.

25. This one is tricky. However, we can view j as a translation of another function. Consider the graph in Figure 9.8. A formula for the graph in Figure 9.8 could be of the form $y = k(x+3)(x+2)(x+1)$. Since $y = 6$ if $x = 0$, $6 = k(0+3)(0+2)(0+1)$, therefore $6 = 6k$, which yields $k = 1$. Note that the graph of $j(x)$ is a vertical shift (by 4) of the graph in Figure 9.8, giving $j(x) = (x+3)(x+2)(x+1) + 4$ as a possible formula for j.

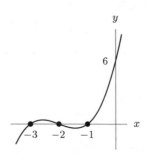

Figure 9.8

29. Clearly $f(x) = x$ works. However, the solution is not unique. If f is of the form $f(x) = ax^2 + bx + c$, then $f(0) = 0$ gives $c = 0$, and $f(1) = 1$ gives

$$a(1)^2 + b(1) + 0 = 1,$$

so

$$a + b = 1,$$

or

$$b = 1 - a.$$

Since these are the only conditions which must be satisfied, any polynomial of the form

$$f(x) = ax^2 + (1-a)x$$

will work. If $a = 0$, we get $f(x) = x$.

33. (a) We could think of $f(x) = (x-2)^3 + 4$ as $y = x^3$ shifted right 2 units and up 4. Thus, since $y = x^3$ is invertible, f should also be. Algebraically, we let

$$y = f(x) = (x-2)^3 + 4.$$

Thus,

$$y - 4 = (x-2)^3$$
$$\sqrt[3]{y-4} = x - 2$$
$$\sqrt[3]{y-4} + 2 = x$$

So $f(x)$ is invertible with an inverse

$$f^{-1}(x) = \sqrt[3]{x-4} + 2.$$

(b) Since g is not so obvious, we might begin by graphing $y = g(x)$. Figure 9.9 shows that the function $g(x) = x^3 - 4x^2 + 2$ does not satisfy the horizontal-line-test, so g is not invertible.

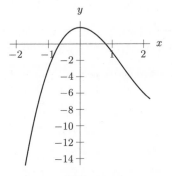

Figure 9.9

37. $y = 4x^2 + 1 = 0$ implies that $x^2 = -\frac{1}{4}$, which has no solutions. There are no real zeros.

41. Let V be the amount of packing material you will need, then

$$V = (\text{Volume of crate}) - (\text{Volume of box})$$
$$= V_c - V_b$$

where V_c and V_b are the volumes of the crate and box, respectively. We have for the box's volume,

$$V_b = \underbrace{\text{length}}_{x} \cdot \overbrace{\text{width}}^{x+2} \cdot \underbrace{\text{depth}}_{x-1} = x(x+2)(x-1).$$

The wooden crate must be 1 ft longer than the cardboard box, so its length is $(x+1)$. This gives the required 0.5-ft clearance between the crate and the front and back of the box. Similarly, the crate's width must be 1 ft greater than the box's width of $(x+2)$, and its depth must be 2 ft greater than the box's depth of $(x-1)$. See Figure 9.10.

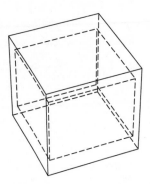

Figure 9.10: Packing a box inside a crate

We have for the crate's volume

$$V_c = \underbrace{\text{length}}_{x+1} \cdot \overbrace{\text{width}}^{(x+2)+1} \cdot \underbrace{\text{depth}}_{(x-1)+2} = (x+1)(x+3)(x+1).$$

Thus, the total amount of packing material will be

$$V = V_c - V_b$$
$$= (x+1)(x+3)(x+1) - x(x+2)(x-1).$$

The formula for V is a difference of two third-degree polynomials. The format is not terribly convenient, so we simplify the formula by multiplying the factors for V_b and V_c and gathering like terms. Then for V we have

$$V = (x+1)(x+3)(x+1) - x(x+2)(x-1)$$
$$= (x^3 + 5x^2 + 7x + 3) - (x^3 + x^2 - 2x)$$
$$= 4x^2 + 9x + 3.$$

The formula $V(x) = 4x^2 + 9x + 3$ gives the necessary information for *appropriate values* of x. Note that the quadratic function $y = 4x^2 + 9x + 3$ is defined for all values of x. However, since x represents the length of a box and $(x-1)$ is the depth of the box, the formula only makes sense as a model for $x > 1$. In this case, an understanding of the component polynomials representing V_b and V_c is necessary in order to determine the logical domain for $V(x)$.

45. (a) We could let $f(x) = k(x+2)(x-3)(x-5)$ to obtain the given zeros. For a y-intercept of 4, $f(0) = 4 = k(0+2)(0-3)(0-5) = 30k$. Thus $30k = 4$, so $k = \frac{2}{15}$. One possible formula is

$$f(x) = \frac{2}{15}(x+2)(x-3)(x-5).$$

(b) One possibility is that f looks like the function in Figure 9.11 and has a double zero at $x = 5$.

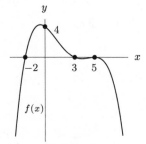

Figure 9.11

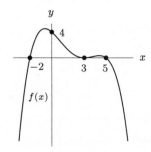

Figure 9.12

Then a formula for f is

$$f(x) = k(x + 2)(x - 3)(x - 5)^2$$

and

$$f(0) = k(2)(-3)(-5)^2.$$

Thus,

$$-150k = 4$$

which gives us

$$k = -\frac{2}{75}.$$

Thus

$$f(x) = -\frac{2}{75}(x + 2)(x - 3)(x - 5)^2.$$

Another possibility is that f has a double-zero at $x = 3$ instead of at $x = 5$. In this case f looks like the function in Figure 9.12. This gives the formula

$$f(x) = -\frac{2}{45}(x + 2)(x - 3)^2(x - 5).$$

Note that if f had a double zero at $x = -2$, there must be another zero for $-2 < x < 0$ in order for f to satisfy $f(0) = 4$ and $y \to -\infty$ as $x \to \pm\infty$.

(c) One possibility is that f looks like the graph in Figure 9.13, with a double zero at $x = -2$ and single zeros at $x = 3$ and $x = 5$.

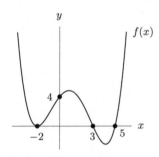

Figure 9.13

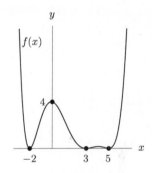

Figure 9.14

A formula for f is $f(x) = k(x + 2)^2(x - 3)(x - 5)$, which gives us

$$k = \frac{1}{15}.$$

Thus,

$$f(x) = \frac{1}{15}(x + 2)^2(x - 3)(x - 5).$$

It is also possible that f has 3 double-zeros at $x = -2$, $x = 3$ and $x = 5$. This leads to a 6th degree polynomial which looks like Figure 9.14. This gives the formula

$$f(x) = \frac{1}{225}(x + 2)^2(x - 3)^2(x - 5)^2.$$

Solutions for Section 9.4

Exercises

1. This is a rational function, as we can put it in the form of one polynomial divided by another:

$$f(x) = \frac{x^2}{2} + \frac{1}{x} = \frac{x^3}{2x} + \frac{2}{2x} = \frac{x^3 + 2}{2x}.$$

5. This is a rational function, as we can put it in the form of one polynomial divided by another:

$$f(x) = \frac{x^2}{x-3} - \frac{5}{x-3} = \frac{x^2 - 5}{x-3}.$$

Problems

9. As $x \to \pm\infty$, $1/x \to 0$, so $f(x) \to 1$. Therefore $y = 1$ is the horizontal asymptote.

13. Let $y = f(x)$. Then $x = f^{-1}(y)$. Solving for x,

$$y = \frac{4 - 3x}{5x - 4}$$
$$y(5x - 4) = 4 - 3x$$
$$5xy - 4y = 4 - 3x$$
$$5xy + 3x = 4y + 4$$
$$x(5y + 3) = 4y + 4 \quad \text{(factor out an } x\text{)}$$
$$x = \frac{4y + 4}{5y + 3}$$

Therefore,

$$f^{-1}(x) = \frac{4x + 4}{5x + 3}.$$

17. (a) From Figure 9.15, we see that

$$\text{Slope of line } l = \frac{\Delta y}{\Delta x} = \frac{C(n_0) - 0}{n_0 - 0} = \frac{C(n_0)}{n_0}.$$

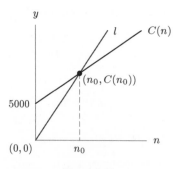

Figure 9.15

(b) $a(n_0) = C(n_0)/n_0$. Thus, the slope of line l is the same as the average cost of producing n_0 units.

21. (a) The formula for h is

$$h(x) = \frac{1}{3}\left(x + x + \frac{2}{x^2}\right) = \frac{1}{3}\left(2x + \frac{2}{x^2}\right) = \frac{2x^3 + 2}{3x^2}.$$

$h(x)$ is the average value of three quantities: x, x, and $2/x^2$. Because one of the quantities, x, is repeated, it is said to be *weighted* in comparison to the other.

(b) As we saw in Problem 20, given an initial guess of $x = 1.26$, it takes about 6 iterations of g to approximate $\sqrt[3]{2}$ to five digits of accuracy. However, since $h(1.26) \approx 1.259921055$, it takes only 1 iteration of h to approximate $\sqrt[3]{2}$ to *eight* digits of accuracy.

Solutions for Section 9.5

Exercises

1. The x-intercept is $x = 2$; the y-intercept is $y = -2/(-4) = 1/2$; the horizontal asymptote is $y = 1$; the vertical asymptote is $x = 4$.

5. The zero of this function is at $x = -3$. It has a vertical asymptote at $x = -5$. Its long-run behavior is: $y \to 1$ as $x \to \pm\infty$. See Figure 9.16.

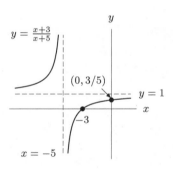

Figure 9.16

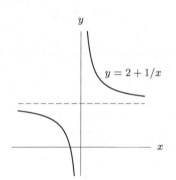

Figure 9.17

9. The graph is the graph of $y = 1/x$ moved up by 2. See Figure 9.17.

13. (a) See the following table.

x	-1	-0.1	-0.01	0	0.01	0.1	1
$F(x)$	0	-99	-9999	Undef	-9999	-99	0

As x approaches 0 from the left the function takes on very large negative values. As x approaches 0 from the right the function takes on very large negative values.

(b)

x	5	10	100	1000
$F(x)$	0.96	0.99	0.9999	0.999999

x	-5	-10	-100	-1000
$F(x)$	0.96	0.99	0.9999	0.999999

For $x > 0$, as x increases, $F(x)$ approaches 1 from below. For $x < 0$, as x decreases, $F(x)$ approaches 1 from below.

(c) The horizontal asymptote is $y = 1$. The vertical asymptote is $x = 0$ (the y-axis). See Figure 9.18.

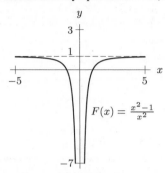

Figure 9.18

Problems

17. (a) If $f(n)$ is large, then $\frac{1}{f(n)}$ is small.
   (b) If $f(n)$ is small, then $\frac{1}{f(n)}$ is large.
   (c) If $f(n) = 0$, then $\frac{1}{f(n)}$ is undefined.
   (d) If $f(n)$ is positive, then $\frac{1}{f(n)}$ is also positive.
   (e) If $f(n)$ is negative, then $\frac{1}{f(n)}$ is negative.

21. The function g is a transformation of $y = 1/x^2$, so $p = 2$. The graph of $y = 1/x^2$ has been shifted 2 units to the right, flipped over the x-axis and shifted 3 units down. To find the y-intercept, we need to evaluate $g(0)$:

$$g(0) = -\frac{1}{4} - 3 = -\frac{13}{4}.$$

Note that this function has no x-intercepts.
   The graph of g is shown in Figure 9.19.

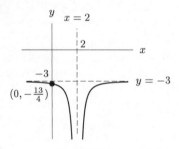

Figure 9.19

25. (a) The graph shows $y = 1/x$ shifted to the right one and up 2 units. Thus,

$$y = \frac{1}{x - 1} + 2$$

is a choice for a formula.
   (b) The equation $y = 1/(x - 1) + 2$ can be written as

$$y = \frac{2x - 1}{x - 1}.$$

   (c) We see that the graph has both an x-and y-intercept. When $x = 0$, $y = \frac{-1}{-1} = 1$, so the y-intercept is $(0, 1)$. If $y = 0$ then $2x - 1 = 0$, so $x = \frac{1}{2}$. The x-intercept is $\left(\frac{1}{2}, 0\right)$.

29. **(a)** The table shows symmetry about the vertical asymptote $x = 3$. The fact that the function values have the same sign on both sides of the vertical asymptotes indicates a transformation of $y = 1/x^2$ rather than $y = 1/x$.

(b) In order to shift the vertical asymptote from $x = 0$ to $x = 3$ for the table, we try

$$y = \frac{1}{(x-3)^2}.$$

checking the x-values from the table in this formula gives y-values that are each 1 less than the y-values of the table. Therefore, we try

$$y = \frac{1}{(x-3)^2} + 1.$$

This formula works. To express the formula as a ratio of polynomials, we take

$$y = \frac{1}{(x-3)^2} + \frac{1(x-3)^2}{(x-3)^2},$$

so

$$y = \frac{x^2 - 6x + 10}{x^2 - 6x + 9}.$$

33. We express the ratio of the volume to the surface area of the box. See Figure 9.20.

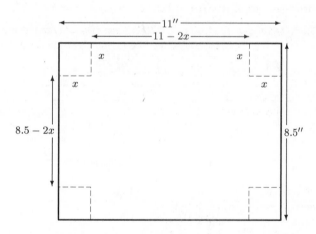

Figure 9.20

The sides of the base are $(11 - 2x)$ and $(8.5 - 2x)$ inches and the depth is x inches, so the volume $V(x) = x(11 - 2x)(8.5 - 2x)$. The surface area, $S(x)$, is the area of the complete sheet minus the area of the four squares, so $S(x) = 11 \cdot 8.5 - 4x^2$. Thus, the ratio we want to maximize is given by

$$R(x) = \frac{V(x)}{S(x)} = \frac{x(11 - 2x)(8.5 - 2x)}{11 \cdot 8.5 - 4x^2}.$$

The graph in Figure 9.21 suggests that the maximum of $R(x)$ occurs when $x \approx 1.939$ inches. (A good viewing window is $0 < x < 5$ and $0 < y < 1$.) So one side is $x = 1.939$ and therefore the others are $11 - 2(1.939) = 7.123$ and $8.5 - 2(1.939) = 4.623$ inches.

The dimensions of the box are 7.123 by 4.623 by 1.939 inches.

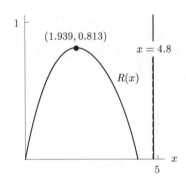

Figure 9.21

37. The graph of $y = \dfrac{x}{(x+2)(x-3)}$ fits.

41. The graph has vertical asymptotes at $x = -1$ and $x = 1$. When $x = 0$, we have $y = 2$ and $y = 0$ at $x = 2$. The graph of $y = \dfrac{(x-2)}{(x+1)(x-1)}$ satisfies each of the requirements, including $y \to 0$ as $x \to \pm\infty$.

45. Factoring the numerator, we have

$$f(x) = \frac{18 - 11x + x^2}{x - 2} = \frac{(x-9)(x-2)}{x-2} = (x-9)\frac{(x-2)}{(x-2)}.$$

In this form we see that the graph of $y = f(x)$ is identical to that of $y = x - 9$, except that the graph of $y = f(x)$ has no y-value corresponding to $x = 2$. The line $y = x - 9$ goes through the point $(2, -7)$ so the graph of $y = f(x)$ will be the line $y = x - 9$ with a hole at $(2, -7)$.

Solutions for Section 9.6

Exercises

1. (a) The function fits neither form. If the expression in the parentheses expanded, then $m(x) = 3(9x^2 + 6x + 1) = 27x^2 + 18x + 3$.
  (b) The function is exponential, because $n(x) = 3 \cdot 2^{3x+1} = 3 \cdot 2^{3x} \cdot 2^1 = 6 \cdot 8^x$.
  (c) The function is exponential, because $p(x) = (5^x)^2 = 5^{2x} = (5^2)^x = 25^x$.
  (d) The function fits neither, because the variable in the exponent is squared.
  (e) The function fits an exponential, because $r(x) = 2 \cdot 3^{-2x} = 2(3^{-2})^x = 2(\frac{1}{9})^x$.
  (f) The function is a power function, because $s(x) = \frac{4}{5x^{-3}} = \frac{4}{5}x^3$.

5. As $x \to \infty$, the higher power dominates, so $x^{1.1}$ dominates $x^{1.08}$. The coefficients 1000 and 50 do not change this, so $y = 50x^{1.1}$ dominates.

Problems

9. Table 9.1 shows that 3^{-x} approaches zero faster than x^{-3} as $x \to \infty$.

Table 9.1

x	2	10	100
3^{-x}	1/9	0.000017	1.94×10^{-48}
x^{-3}	1/8	0.001	10^{-6}

13. (a) If f is linear, then the formula for $f(x)$ is of the form

$$f(x) = mx + b,$$

where

$$m = \frac{\Delta y}{\Delta x} = \frac{48 - \frac{3}{4}}{2 - (-1)} = \frac{\frac{189}{4}}{3} = \frac{189}{12} = \frac{63}{4}$$

Thus, $f(x) = \frac{63}{4}x + b$. Since $f(2) = 48$, we have

$$48 = \frac{63}{4}(2) + b$$

$$48 - \frac{63}{2} = b$$

$$b = \frac{33}{2}$$

Thus, if f is a linear function,

$$f(x) = \frac{63}{4}x + \frac{33}{2}.$$

(b) If f is exponential, then the formula for $f(x)$ is of the form

$$f(x) = ab^x, \qquad b > 0,\ b \neq 1.$$

Taking the ratio of $f(2)$ to $f(-1)$, we have

$$\frac{f(2)}{f(-1)} = \frac{ab^2}{ab^{-1}} = b^3,$$

and

$$\frac{f(2)}{f(-1)} = \frac{48}{\frac{3}{4}} = 48 \cdot \frac{4}{3} = 64.$$

Thus

$$b^3 = 64,$$

and

$$b = 4.$$

To solve for a, note that

$$f(2) = a(4)^2 = 48,$$

which gives $a = 3$. Thus, an exponential model for f is $f(x) = 3 \cdot 4^x$.

(c) If f is a power function, then the formula for $f(x)$ is of the form

$$f(x) = kx^p, \qquad k \text{ and } p \text{ constants.}$$

Taking the ratio of $f(2)$ to $f(-1)$, we have

$$\frac{f(2)}{f(-1)} = \frac{k \cdot 2^p}{k \cdot (-1)^p} = \frac{2^p}{(-1)^p} = (-2)^p.$$

Since we know from part (b) that $\frac{f(2)}{f(-1)} = 64$, we have

$$(-2)^p = 64.$$

Thus, $p = 6$. To solve for k, note that

$$f(2) = k \cdot 2^6 = 48,$$

which gives

$$64k = 48$$

$$k = \frac{48}{64} = \frac{3}{4}.$$

Thus, $f(x) = \frac{3}{4}x^6$ is a power function which satisfies the given data.

17. If $f(x) = mx^{1/3}$ goes through $(1, 2)$, then $m(1)^{1/3} = 2$, so $m = 2$ and $f(x) = 2x^{1/3}$. Using $x = 8$ in $f(x) = 2x^{1/3}$ gives $t = 4$. If $g(x) = kx^{4/3}$ goes through $(8, 4)$, then $k = \frac{1}{4}$. Thus, $m = 2, t = 4$, and $k = \frac{1}{4}$.

21. Multiplying out the numerator gives a polynomial with highest term x^3, which dominates the x^2 in the denominator. Thus, $y \to x$ as $x \to \infty$ or $x \to -\infty$. So $y \to \infty$ as $x \to \infty$, and $y \to -\infty$ as $x \to -\infty$.

25. Since e^x dominates x^{100}, for large positive x, the value of y is very large. Thus, $y \to \infty$ as $x \to \infty$.
   For large negative x, the value of e^x is very small, but x^{100} is very large. Thus, $y \to 0$ as $x \to -\infty$.

29. As $x \to \infty$, the value of $e^x \to \infty$ and $e^{-x} \to 0$. As $x \to -\infty$, the value of $e^{-x} \to \infty$ and $e^x \to 0$. Thus, $y \to \infty$ as $x \to \infty$ and $y \to -\infty$ as $x \to -\infty$.

33. We know that $p = 365 \left(\frac{d}{93} \right)^{3/2}$. Then by substituting $d = \frac{1}{2}(93)$, we get $p = 365 \left(\frac{1}{2} \right)^{3/2} \approx 129$ Earth days.

37. The function $f(d) = b \cdot d^{p/q}$, with $p < q$, because $f(d)$ increases more and more slowly as d gets larger, and $g(d) = a \cdot d^{p/q}$, with $p > q$, because $g(d)$ increases more and more quickly as d gets larger.

Solutions for Section 9.7

Exercises

1. $f(x) = ax^p$ for some constants a and c. Since $f(1) = 1 = a(1)^p$, it follows that $a = 1$. Also, $f(2) = 2^p = c$. Solving for p we have $p = \ln c / \ln 2$. Thus, $f(x) = x^{\ln c / \ln 2}$

5. **(a)** Regression on a calculator returns the power function $f(x) = 201.353x^{2.111}$, where $f(x)$ represents the total dry weight (in grams) of a tree having an x cm diameter at breast height.
   (b) Using our regression function, we obtain $f(20) = 201.535(20)^{2.111} = 112,313.62$ gm.
   (c) Solving $f(x) = 201.353x^{2.111} = 100,000$ for x we get

$$x^{2.111} = \frac{100,000}{201.353}$$

$$x = \left(\frac{100,000}{201.353} \right)^{1/2.111} = 18.930 \text{ cm.}$$

Problems

9. **(a)**

Recliner price ($)	399	499	599	699	799
Demand (recliners)	62	55	47	40	34
Revenue ($)	24,738	27,445	28,153	27,960	27,166

   (b) Using quadratic regression, we obtain the following formula for revenue, R, as a function of the selling price, p:
   $R(p) = -0.0565p^2 + 72.9981p + 4749.85$.
   (c) By using a graphing calculator to zoom in, we see that the price which maximizes revenue is about $646. The revenue generated at this price is about $28,349.

13. **(a)** Quadratic is the only choice that increases and then decreases to match the data.
   (b) Using ages of $x = 20, 30, \ldots, 80$, a quadratic function is $y = -0.016x^2 + 1.913x - 10$. Answers may vary.
   (c) The value of the function at 37 is $y = -0.016 \cdot 37^2 + 1.913 \cdot 37 - 10 = \$38,877$.
   (d) The value of the function for age 10 is $y = -0.016 \cdot 10^2 + 1.913 \cdot 10 - 10 = \$7,530$. Not reasonable, ten-year-olds don't usually work. Answers may vary.

17. **(a)** Table 9.2 shows the transformed data, where $y = \ln N$. Figure 9.22 shows that the transformed data lie close to a straight line.

$$y = \ln N$$

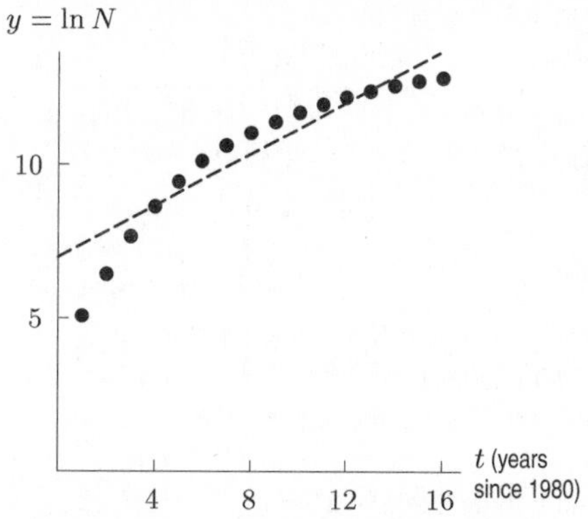

Figure 9.22: Domestic deaths from AIDS,
1981–96 ($\ln N$ against t)

Table 9.2

t	$y = \ln N$	t	$y = \ln N$
1	5.069	9	11.408
2	6.433	10	11.708
3	7.664	11	11.972
4	8.637	12	12.203
5	9.442	13	12.405
6	10.115	14	12.587
7	10.624	15	12.740
8	11.039	16	12.837

(b) We now use linear regression to estimate a line to fit the data points $(t, \ln N)$. The formula provided by a calculator (and rounded) is

$$y = 6.445 + 0.47t.$$

(c) To find the formula for N in terms of t, we substitute $\ln N$ for y and solve for N:

$$\ln N = 6.445 + 0.47t$$
$$N = e^{6.445 + 0.47t}$$
$$= \left(e^{6.445}\right)\left(e^{0.47t}\right),$$

and since $e^{6.445} \approx 630$, we have

$$N \approx 630 e^{0.47t}.$$

21. (a) $y = 2x$

(b) $\ln y = 2 \ln x$, so $\ln y = \ln\left(x^2\right)$, and $y = x^2$.

(c) $\ln y = 2x$, so $y = e^{2x}$.

Solutions for Chapter 9 Review

Exercises

1. Since the graph is symmetric about the y-axis, the power function is even.

5. By multiplying out the expression $x(x - 3)(x + 2)$ and then simplifying the result, we see that

$$u(x) = x^3 - x^2 - 6x,$$

So u is a third-degree polynomial.

9. Zeros occur where $y = 0$, which we can find by factoring:

$$(x^2 - 8x + 12)(x - 3) = y$$
$$(x - 6)(x - 2)(x - 3) = y.$$

Zeros are at $x = 6$, $x = 2$, and $x = 3$.

13. Since $y = 7/x^{-4} = 7x^4$ and since x^4 dominates x^3, we see that $7/x^{-4}$ dominates.

Problems

17. Try $f(x) = k(x+1)(x-1)^2$ because f has a zero at $x = -1$ and a double zero at $x = 1$. Since $f(0) = -1$, we have $f(0) = k(0+1)(0-1)^2 = k$; thus $k = -1$. So

$$f(x) = -(x+1)(x-1)^2.$$

21. Notice that we can think of g as a vertically shifted polynomial. That is, if we let $g(x) = h(x) + 4$, then $h(x)$ is a polynomial with zeros at $x = -1$, $x = 2$, and $x = 4$; furthermore, since $g(-2) = 0$, $h(-2) = 0 - 4 = -4$. Thus,

$$h(x) = k(x+1)(x-2)(x-4).$$

To find k, note that $h(-2) = k(-2+1)(-2-2)(-2-4) = k(-1)(-4)(-6) = -24k$. Since $h(-2) = -24k = -4$, we have $k = \frac{1}{6}$, which gives

$$h(x) = \frac{1}{6}(x+1)(x-2)(x-4).$$

Thus since $g(x) = h(x) + 4$, we have

$$g(x) = \frac{1}{6}(x+1)(x-2)(x-4) + 4.$$

25. (a) If f is even, then for all x

$$f(x) = f(-x).$$

Since $f(x) = ax^2 + bx + c$, this implies

$$ax^2 + bx + c = a(-x)^2 + b(-x) + c$$
$$= ax^2 - bx + c$$

We can cancel the ax^2 term and the constant term c from both sides of this equation, giving

$$bx = -bx$$
$$bx + bx = 0$$
$$2bx = 0.$$

Since x is not necessarily zero, we conclude that b must equal zero, so that if f is even,

$$f(x) = ax^2 + c.$$

(b) If g is odd, then for all x

$$g(-x) = -g(x).$$

Since $g(x) = ax^3 + bx^2 + cx + d$, this implies

$$ax^3 + bx^2 + cx + d = -(a(-x)^3 + b(-x)^2 + c(-x) + d)$$
$$= -(-ax^3 + bx^2 - cx + d)$$
$$= ax^3 - bx^2 + cx - d.$$

The odd-powered terms cancel, leaving

$$bx^2 + d = -bx^2 - d$$
$$2bx^2 + 2d = 0$$
$$bx^2 + d = 0.$$

Since this must hold true for any value of x, we know that both b and d must equal zero. Therefore,

$$g(x) = ax^3 + cx.$$

29. (a) $f(x) = \dfrac{\text{Amount of Alcohol}}{\text{Amount of Liquid}} = \dfrac{x}{x+5}$

(b) $f(7) = \frac{7}{7+5} = \frac{7}{12} \approx 58.333\%$. Also, $f(7)$ is the concentration of alcohol in a solution consisting of 5 gallons of water and 7 gallons of alcohol.

(c) $f(x) = 0$ implies that $\dfrac{x}{x+5} = 0$ and so $x = 0$. The concentration of alcohol is 0% when there is no alcohol in the solution, that is, when $x = 0$.

(d) The horizontal asymptote is given by the ratio of the highest-power terms of the numerator and denominator:

$$y = \frac{x}{x} = 1 = 100\%$$

This means that as the amount of alcohol added, x, grows large, the concentration of alcohol in the solution approaches 100%.

33. $A < 0 < C < D < B$

37.
$$f(x) = \frac{p(x)}{q(x)} = \frac{-3(x-2)(x-3)}{(x-5)^2}$$

We need the factor of -3 in the numerator and the exponent of 2 in the denominator, because we have a horizontal asymptote of $y = -3$. The ratio of highest term of $p(x)$ to highest term of $q(x)$ will be $\frac{-3x^2}{x^2} = -3$.

41. (a) If $k = 3$, then

$$\phi^k + \phi^{k+1} = \phi^3 + \phi^4$$
$$= \left(\frac{1+\sqrt{5}}{2}\right)^3 + \left(\frac{1+\sqrt{5}}{2}\right)^4$$
$$\approx 4.24 + 6.85 = 11.09.$$

We see that this equals

$$\phi^5 = \left(\frac{1+\sqrt{5}}{2}\right)^5 \approx 11.09.$$

If $k = 10$, then

$$\phi^k + \phi^{k+1} = \phi^{10} + \phi^{11}$$
$$= \left(\frac{1+\sqrt{5}}{2}\right)^{10} + \left(\frac{1+\sqrt{5}}{2}\right)^{11}$$
$$\approx 123 + 199 = 322.$$

We see that this equals

$$\phi^{12} = \left(\frac{1+\sqrt{5}}{2}\right)^{12} \approx 322.$$

(b) We know from the solution to Problem 40 that
$$\phi^2 = \phi + 1.$$

Multiplying both sides of this equation by ϕ^k, we have

$$\phi^k \cdot \phi^2 = \phi^k(\phi + 1)$$
$$\phi^{k+2} = \phi^k \cdot \phi + \phi^k$$
$$\phi^{k+2} = \phi^{k+1} + \phi^k.$$

45. The table shows 3 values of x for which $y = 0$. The function does not appear periodic, so a polynomial may be the best choice. The zeros at $x = -2$, $x = 1$ and $x = 2$ suggest a cubic of the form

$$y = k(x+2)(x-1)(x-2).$$

Since $y = 8$ when $x = 0$,

$$8 = k(2)(-1)(-2)$$
$$k = 2.$$

Note that

$$y = 2(x + 2)(x - 1)(x - 2)$$

fits the data exactly.

49. **(a)** We know that a planet's capture cross-section A is greater than πR^2, which is the apparent area of the planet from a distance. This is because we cannot miss the planet if we aim directly towards it. We can get the same result algebraically. Because v^2, M, G, and R are always positive, $(2MG/R)/v^2$ is always positive, and thus $1 + (2MG/R)/v^2$ is always greater than 1. Therefore,

$$\pi R^2 \left(1 + \frac{2MG/R}{v^2}\right)$$

is always greater than πR^2. We conclude that a planet's capture cross-section exceeds its apparent area due to the effects of gravity.

(b) Let $A_1(v)$ and $A_2(v)$ denote the capture cross-sections of the two planets, respectively. If M and R are the mass and radius of the first planet, then the mass of the second planet is $M/2$, and its radius is $2R$, which means that

$$A_1(v) = \pi R^2 \left(1 + \frac{2MG/R}{v^2}\right)$$

$$\text{and} \quad A_2(v) = \pi(2R)^2 \left(1 + \frac{2(M/2)G/(2R)}{v^2}\right)$$

$$= 4\pi R^2 \left(1 + \frac{MG/2R}{v^2}\right)$$

$$= \pi R^2 \left(4 + \frac{2MG/R}{v^2}\right).$$

If we write

$$A_1(v) = \pi R^2 + \frac{2\pi R^2 G/R}{v^2} \quad \text{and} \quad A_2(v) = 4\pi R^2 + \frac{2\pi R^2 G/R}{v^2},$$

we see that $A_2(v)$ is larger than $A_1(v)$, the difference being exactly $3\pi R^2$.

(c) The equation of the horizontal asymptote is $y = \pi R^2$. This means that as v becomes very large, the planet's capture cross-section approaches πR^2, which is the apparent area of the planet. This makes sense, because in order for a rapidly moving spacecraft to hit a planet, our aim must be very good.

The equation of the vertical asymptote is $v = 0$. This means that as v becomes very small, the planet's capture cross-section becomes very large. This makes sense, because a slowly drifting spacecraft has an excellent chance of being dragged into the planet by its gravitational force, even if its aim was wide of the mark. Note that for a spacecraft initially at rest, $v = 0$ and A is undefined. We interpret this as meaning a spacecraft initially at rest will inevitably strike the planet, whose cross-section is "infinitely" large. This makes sense, because, with no other forces acting on it and no momentum of its own, a spacecraft would begin falling towards the planet due to the effects of gravity.

CHECK YOUR UNDERSTANDING

1. False. The quadratic function $y = 3x^2 + 5$ is not of the form $y = kx^n$, so it is not a power function.

5. True. All positive even power functions have an upward opening U shape.

9. True. The x-axis is an asymptote for $f(x) = x^{-1}$, so the values approach zero.

13. False. As x grows very large the exponential decay function g approaches the x-axis faster than any power function with a negative power.

17. False. For example, the polynomial $x^2 + x^3$ has degree 3 because the degree is the highest power, not the first power, in the formula for the polynomial.

21. True. The graph crosses the y-axis at the point $(0, p(0))$.

25. True. We can write $p(x) = (x - a) \cdot C(x)$. Evaluating at $x = a$ we get $p(a) = (a - a) \cdot C(a) = 0 \cdot C(a) = 0$.

29. True. This is the definition of a rational function.

33. True. The ratio of the highest degree terms in the numerator and denominator is $2x/x^2 = 2/x$, so for large positive x-values, y approaches 0.

37. False. The ratio of the highest degree terms in the numerator and denominator is $3x^4/x^2 = 3x^2$. So for large positive x-values, y behaves like $y = 3x^2$.

41. True. At $x = -4$, we have $f(-4) = (-4 + 4)/(-4 - 3) = 0/(-7) = 0$, so $x = -4$ is a zero.

45. False. If $p(x)$ has no zeros, then $r(x)$ has no zeros. For example, if $p(x)$ is a nonzero constant or $p(x) = x^2 + 1$, then $r(x)$ has no zeros.

Solutions to Tools for Chapter 9

1. $\dfrac{3}{5} + \dfrac{4}{7} = \dfrac{3 \cdot 7 + 4 \cdot 5}{35} = \dfrac{21 + 20}{35} = \dfrac{41}{35}$

5. $\dfrac{-2}{yz} + \dfrac{4}{z} = \dfrac{-2z + 4yz}{yz^2} = \dfrac{-2 + 4y}{yz} = \dfrac{-2(1 - 2y)}{yz}$

9. $\dfrac{\frac{3}{4}}{\frac{7}{20}} = \dfrac{3}{4} \cdot \dfrac{20}{7} = \dfrac{60}{28} = \dfrac{15}{7}$

13. $\dfrac{13}{x - 1} + \dfrac{14}{2x - 2} = \dfrac{13}{x - 1} + \dfrac{14}{2(x - 1)} = \dfrac{13 \cdot 2 + 14}{2(x - 1)} = \dfrac{40}{2(x - 1)} = \dfrac{20}{x - 1}$

17. $\dfrac{8y}{y - 4} + \dfrac{32}{y - 4} = \dfrac{8y + 32}{y - 4} = \dfrac{8(y + 4)}{y - 4}$

21.

$$\dfrac{8}{3x^2 - x - 4} - \dfrac{9}{x + 1} = \dfrac{8}{(x + 1)(3x - 4)} - \dfrac{9}{x + 1}$$
$$= \dfrac{8 - 9(3x - 4)}{(x + 1)(3x - 4)}$$
$$= \dfrac{-27x + 44}{(x + 1)(3x - 4)}$$

25. If we rewrite the second fraction $-\dfrac{1}{1 - x}$ as $\dfrac{1}{x - 1}$, the common denominator becomes $x - 1$. Therefore,

$$\dfrac{x^2}{x - 1} - \dfrac{1}{1 - x} = \dfrac{x^2}{x - 1} + \dfrac{1}{x - 1} = \dfrac{x^2 + 1}{x - 1}.$$

29. The common denominator is e^{2x}. Thus,

$$\dfrac{1}{e^{2x}} + \dfrac{1}{e^x} = \dfrac{1}{e^{2x}} + \dfrac{e^x}{e^{2x}} = \dfrac{1 + e^x}{e^{2x}}.$$

33. $\dfrac{8y}{y - 4} - \dfrac{32}{y - 4} = \dfrac{8y - 32}{y - 4} = \dfrac{8(y - 4)}{y - 4} = 8$

37. We write this complex fraction as a multiplication problem. Therefore,

$$\frac{\frac{w+2}{2}}{w+2} = \frac{w+2}{2} \cdot \frac{1}{w+2} = \frac{1}{2}.$$

41. We expand within the first brackets first. Therefore,

$$\frac{[4-(x+h)^2]-[4-x^2]}{h} = \frac{[4-(x^2+2xh+h^2)]-[4-x^2]}{h}$$
$$= \frac{[4-x^2-2xh-h^2]-4+x^2}{h} = \frac{-2xh-h^2}{h}$$
$$= -2x-h.$$

45. We simplify the second complex fraction first. Thus,

$$p - \frac{q}{\frac{p}{q}+\frac{q}{p}} = p - \frac{q}{\frac{p^2+q^2}{qp}} = p - q \cdot \frac{qp}{p^2+q^2}$$
$$= \frac{p(p^2+q^2)-q^2p}{p^2+q^2} = \frac{p^3}{p^2+q^2}.$$

49. Write

$$\frac{\frac{1}{2}(2x-1)^{-1/2}(2)-(2x-1)^{1/2}(2x)}{(x^2)^2} = \frac{\frac{1}{(2x-1)^{1/2}} - \frac{2x(2x-1)^{1/2}}{1}}{(x^2)^2}.$$

Next a common denominator for the top two fractions is $(2x-1)^{1/2}$. Therefore we obtain,

$$\frac{\frac{1}{(2x-1)^{1/2}} - \frac{2x(2x-1)}{(2x-1)^{1/2}}}{x^4} = \frac{1-4x^2+2x}{(2x-1)^{1/2}} \cdot \frac{1}{x^4} = \frac{-4x^2+2x+1}{x^4\sqrt{2x-1}}.$$

53. The denominator p^2+11 is divided into each of the two terms of the numerator. Thus,

$$\frac{7+p}{p^2+11} = \frac{7}{p^2+11} + \frac{p}{p^2+11}.$$

57. The numerator $q-1 = q-4+3$. Thus,

$$\frac{q-1}{q-4} = \frac{(q-4)+3}{q-4} = 1 + \frac{3}{q-4}.$$

61.

$$\frac{1+e^x}{e^x} = \frac{1}{e^x} + \frac{e^x}{e^x} = \frac{1}{e^x} + 1 = 1 + \frac{1}{e^x}$$

65. False

CHAPTER TEN

Solutions for Section 10.1

Exercises

1. Scalar.

5.
$$\vec{p} = 2\vec{w}, \quad \vec{q} = -\vec{u}, \quad \vec{r} = \vec{w} + \vec{u} = \vec{u} + \vec{w},$$
$$\vec{s} = \vec{p} + \vec{q} = 2\vec{w} - \vec{u}, \quad \vec{t} = \vec{u} - \vec{w}$$

9. See Figure 10.1.

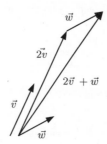

Figure 10.1

Problems

13.

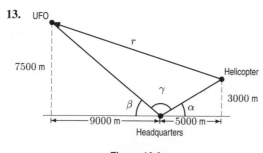

Figure 10.2

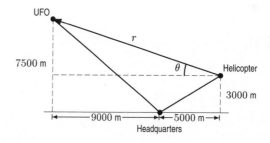

Figure 10.3

Figure 10.2 shows the headquarters at the origin, and a positive y-value as up, and a positive x-value as east. To solve for r, we must first find γ:

$$\gamma = 180° - \alpha - \beta$$
$$= 180° - \arctan\frac{3000}{5000} - \arctan\frac{7500}{9000}$$
$$= 109.231°.$$

We now can find r using the Law of Cosines in the triangle formed by the position of the headquarters, the helicopter and the UFO.

In kilometers:

$$r^2 = 34 + 137.250 - 2 \cdot \sqrt{34} \cdot \sqrt{137.250} \cdot \cos\gamma$$
$$r^2 = 216.250$$
$$r = 14.705 \text{ km}$$
$$= 14,705 \text{ m.}$$

From Figure 10.3 we see:

$$\tan\theta = \frac{4500}{14,000}$$
$$\theta = 17.819°.$$

Therefore, the helicopter must fly 14,705 meters with an angle of $17.819°$ from the horizontal.

17. The effect of scaling the left-hand picture in Figure 10.4 is to stretch each vector by a factor of a (shown with $a > 1$). Since, after scaling up, the three vectors $a\vec{v}$, $a\vec{w}$, and $a(\vec{v} + \vec{w})$ form a similar triangle, we know that $a(\vec{v} + \vec{w})$ is the sum of the other two: that is

$$a(\vec{v} + \vec{w}) = a\vec{v} + a\vec{w} .$$

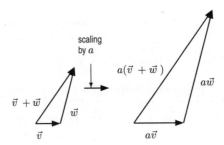

Figure 10.4

21. By Figure 10.5, the vectors $\vec{v} + (-1)\vec{w}$ and $\vec{v} - \vec{w}$ are equal.

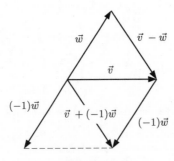

Figure 10.5

Solutions for Section 10.2

Exercises

1. $5(2\vec{i} - \vec{j}) + \vec{j} = 10\vec{i} - 5\vec{j} + \vec{j} = 10\vec{i} - 4\vec{j}$.

5.

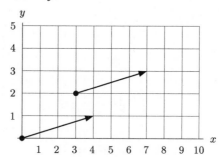

Figure 10.6: $\vec{v}$

9. $\|\vec{v}\| = \sqrt{1^2 + (-1)^2 + 3^2} = \sqrt{11} \approx 3.317$

Problems

13. (a) $\vec{v} = 2\vec{i} + \vec{j}$
   (b) Since $\vec{w} = \vec{i} - \vec{j}$, we have $2\vec{w} = 2\vec{i} - 2\vec{j}$.
   (c) Since $\vec{v} = 2\vec{i} + \vec{j}$ and $\vec{w} = \vec{i} - \vec{j}$, we have $\vec{v} + \vec{w} = (2\vec{i} + \vec{j}) + (\vec{i} - \vec{j}) = 3\vec{i}$.
   (d) Since $\vec{w} = \vec{i} - \vec{j}$ and $\vec{v} = 2\vec{i} + \vec{j}$, we have $\vec{w} - \vec{v} = (\vec{i} - \vec{j}) - (2\vec{i} + \vec{j}) = -\vec{i} - 2\vec{j}$.
   (e) $\overrightarrow{PQ} = \vec{i} + \vec{j}$
   (f) Since P is at the point $(1, -2)$, the vector we want is $(2 - 1)\vec{i} + (0 - (-2))\vec{j} = \vec{i} + 2\vec{j}$.
   (g) The vector must be horizontal, so $\vec{i}$ will work.
   (h) The vector must be vertical, so $\vec{j}$ will work.

17. We need to calculate the length of each vector.

$$\|21\vec{i} + 35\vec{j}\| = \sqrt{21^2 + 35^2} = \sqrt{1666} \approx 40.8,$$
$$\|40\vec{i}\| = \sqrt{40^2} = 40.$$

So the first car is faster.

21. To determine if two vectors are parallel, we need to see if one vector is a scalar multiple of the other one. Since $\vec{u} = -2\vec{w}$, and $\vec{v} = \frac{1}{4}\vec{q}$ and no other pairs have this property, only $\vec{u}$ and $\vec{w}$, and $\vec{v}$ and $\vec{q}$ are parallel.

25.

$$\text{Displacement} = \text{Cat's coordinates} - \text{Bottom of the tree's coordinates}$$
$$= (1 - 2)\vec{i} + (4 - 4)\vec{j} + (0 - 0)\vec{k} = -\vec{i}.$$

Solutions for Section 10.3

Exercises

1. $\vec{G} = \vec{N} + \vec{M} = (5, 6, 7, 8, 9, 10) + (1, 1, 2, 3, 5, 8) = (6, 7, 9, 11, 14, 18)$.

5. $\vec{K} = \dfrac{\vec{N}}{3} + \dfrac{2\vec{N}}{3} = \dfrac{3\vec{N}}{3} = \vec{N} = (5, 6, 7, 8, 9, 10).$

9. Since each population decreases by 22%, we multiply $\vec{Q}$ by $1 - 0.22 = 0.78$ to find $\vec{T}$:

$$\vec{T} = 0.78\vec{Q} = 0.78(3.43, 1.29, 6.38, 1.26, 1.06, 0.61) = (2.675, 1.006, 4.976, 0.983, 0.827, 0.476).$$

Problems

13. (a)

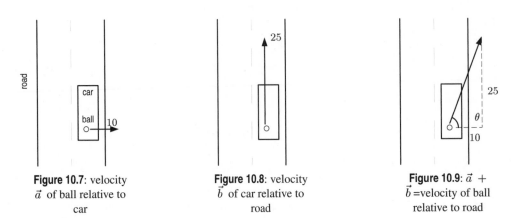

Figure 10.7: velocity $\vec{a}$ of ball relative to car

Figure 10.8: velocity $\vec{b}$ of car relative to road

Figure 10.9: $\vec{a} + \vec{b}$ =velocity of ball relative to road

(b) Solving $\tan \theta = 25/10$ gives $\theta = 68.199°$. So the angle the ball makes with the road is $90° - 68.199° = 21.801°$.

17. Let the x-axis point east and the y-axis point north. Since the wind is blowing from the northeast at a speed of 50 km/hr, the velocity of the wind is

$$\vec{w} = -50 \cos 45°\vec{i} - 50 \sin 45°\vec{j} \approx -35.355\vec{i} - 35.355\vec{j}.$$

Let $\vec{a}$ be the velocity of the airplane, relative to the air, and let ϕ be the angle from the x-axis to $\vec{a}$; since $\|\vec{a}\| = 600$ km/hr, we have $\vec{a} = 600 \cos \phi\vec{i} + 600 \sin \phi\vec{j}$. (See Figure 10.10.)

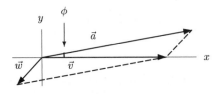

Figure 10.10

Now the resultant velocity, $\vec{v}$, is given by

$$\vec{v} = \vec{a} + \vec{w} = (600 \cos \phi\vec{i} + 600 \sin \phi\vec{j}) + (-35.355\vec{i} - 35.355\vec{j})$$
$$= (600 \cos \phi - 35.355)\vec{i} + (600 \sin \phi - 35.355)\vec{j}.$$

Since the airplane is to fly due east, i.e., in the x direction, then the y-component of the velocity must be 0, so we must have

$$600 \sin \phi - 35.355 = 0$$
$$\sin \phi = \frac{35.355}{600}.$$

Thus $\phi = \arcsin(35.355/600) \approx 3.378°$.

Solutions for Section 10.4

Exercises

1. $\vec{c} \cdot \vec{y} = (\vec{i} + 6\vec{j}) \cdot (4\vec{i} - 7\vec{j}) = 1 \cdot 4 + 6(-7) = 4 - 42 = -38.$

5. $\vec{b} \cdot \vec{z} = (-3\vec{i} + 5\vec{j} + 4\vec{k}) \cdot (\vec{i} - 3\vec{j} - \vec{k}) = -22.$

9. Since $\vec{a} \cdot \vec{y}$ and $\vec{c} \cdot \vec{z}$ are both scalars, the expression is the product of two numbers and therefore a number. We have

$$\vec{a} \cdot \vec{y} = (2\vec{j} + \vec{k}) \cdot (4\vec{i} - 7\vec{j}) = 0(4) + 2(-7) + 1(0) = -14$$

$$\vec{c} \cdot \vec{z} = (\vec{i} + 6\vec{j}) \cdot (\vec{i} - 3\vec{j} - \vec{k}) = 1(1) + 6(-3) + 0(-1) = -17.$$

Thus,

$$(\vec{a} \cdot \vec{y})(\vec{c} \cdot \vec{z}) = 238.$$

Problems

13. Since $3\vec{i} + \sqrt{3}\vec{j} = \sqrt{3}(\sqrt{3}\vec{i} + \vec{j})$, we know that $3\vec{i} + \sqrt{3}\vec{j}$ and $\sqrt{3}\vec{i} + \vec{j}$ are scalar multiples of one another, and therefore parallel.

Since $(\sqrt{3}\vec{i} + \vec{j}) \cdot (\vec{i} - \sqrt{3}\vec{j}) = \sqrt{3} - \sqrt{3} = 0$, we know that $\sqrt{3}\vec{i} + \vec{j}$ and $\vec{i} - \sqrt{3}\vec{j}$ are perpendicular.

Since $3\vec{i} + \sqrt{3}\vec{j}$ and $\sqrt{3}\vec{i} + \vec{j}$ are parallel, $3\vec{i} + \sqrt{3}\vec{j}$ and $\vec{i} - \sqrt{3}\vec{j}$ are perpendicular, too.

17. We need to find the speed of the wind in the direction of the track. Looking at Figure 10.11, we see that we want the component of $\vec{w}$ in the direction of $\vec{v}$. We calculate

$$\|\vec{w}_{\text{parallel to } \vec{v}}\| = \|\vec{w}\| \cos\theta = \frac{\vec{w} \cdot \vec{v}}{\|\vec{v}\|} = \frac{(5\vec{i} + \vec{j}) \cdot (2\vec{i} + 6\vec{j})}{\|2\vec{i} + 6\vec{j}\|}$$

$$= \frac{16}{\sqrt{40}} \approx 2.53 < 5.$$

Therefore, the race results will not be disqualified.

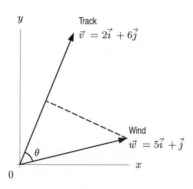

Figure 10.11

21. By the distributive law,

$$\vec{u} \cdot \vec{v} = (u_1\vec{i} + u_2\vec{j}) \cdot (v_1\vec{i} + v_2\vec{j})$$
$$= u_1 v_1 \vec{i} \cdot \vec{i} + u_1 v_2 \vec{i} \cdot \vec{j} + u_2 v_1 \vec{j} \cdot \vec{i} + u_2 v_2 \vec{j} \cdot \vec{j}.$$

Since $\|\vec{i}\| = \|\vec{j}\| = 1$ and the angle between $\vec{i}$ and $\vec{j}$ is $90°$,

$$\vec{i} \cdot \vec{i} = 1, \quad \vec{j} \cdot \vec{j} = 1, \quad \vec{i} \cdot \vec{j} = \vec{j} \cdot \vec{i} = 0.$$

Thus,

$$\vec{u} \cdot \vec{v} = u_1 v_1 + u_2 v_2.$$

25. Let the room be put in the coordinate system as shown in Figure 10.12.

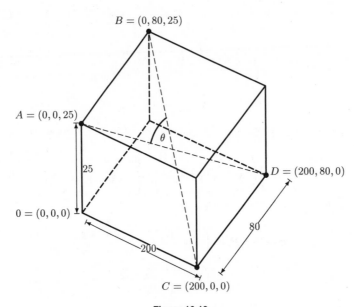

Figure 10.12

Then the vectors of the two strings are given by:
$$\overrightarrow{AD} = (200\vec{i} + 80\vec{j} + 0\vec{k}) - (0\vec{i} + 0\vec{j} + 25\vec{k}) = 200\vec{i} + 80\vec{j} - 25\vec{k}$$
$$\overrightarrow{BC} = (200\vec{i} + 0\vec{j} + 0\vec{k}) - (0\vec{i} + 80\vec{j} + 25\vec{k}) = 200\vec{i} - 80\vec{j} - 25\vec{k}.$$
Let the angle between $\overrightarrow{AD}$ and $\overrightarrow{BC}$ be θ. Then we have

$$\begin{aligned}
\cos\theta &= \frac{\overrightarrow{AD} \cdot \overrightarrow{BC}}{\|\overrightarrow{AD}\| \|\overrightarrow{BC}\|} \\
&= \frac{200(200) + (80)(-80) + (-25)(-25)}{\sqrt{200^2 + 80^2 + (-25)^2}\sqrt{(200)^2 + (-80)^2 + (-25)^2}} \\
&= \frac{34225}{47025} \\
&= 0.727804 \\
\theta &= \arccos 0.727804 \\
&= 43.297°.
\end{aligned}$$

Solutions for Section 10.5

Problems

1. (a) We have

$$5\mathbf{R} = 5\begin{pmatrix} 3 & 7 \\ 2 & -1 \end{pmatrix} = \begin{pmatrix} 5\cdot 3 & 5\cdot 7 \\ 5\cdot 2 & 5\cdot -1 \end{pmatrix} = \begin{pmatrix} 15 & 35 \\ 10 & -5 \end{pmatrix}.$$

(b) We have

$$-2\mathbf{S} = -2\begin{pmatrix} 1 & -5 \\ 0 & 8 \end{pmatrix} = \begin{pmatrix} -2\cdot 1 & -2\cdot -5 \\ -2\cdot 0 & -2\cdot 8 \end{pmatrix} = \begin{pmatrix} -2 & 10 \\ 0 & -16 \end{pmatrix}.$$

(c) We have

$$\mathbf{R} + \mathbf{S} = \begin{pmatrix} 3 & 7 \\ 2 & -1 \end{pmatrix} + \begin{pmatrix} 1 & -5 \\ 0 & 8 \end{pmatrix}$$

$$= \begin{pmatrix} 3+1 & 7-5 \\ 2+0 & -1+8 \end{pmatrix} = \begin{pmatrix} 4 & 2 \\ 2 & 7 \end{pmatrix}.$$

(d) Writing $\mathbf{S} - 3\mathbf{R} = \mathbf{S} + (-3)\mathbf{R}$, we first find $-3\mathbf{R}$:

$$-3\mathbf{R} = -3\begin{pmatrix} 3 & 7 \\ 2 & -1 \end{pmatrix} = \begin{pmatrix} -3\cdot 3 & -3\cdot 7 \\ -3\cdot 2 & -3\cdot -1 \end{pmatrix} = \begin{pmatrix} -9 & -21 \\ -6 & 3 \end{pmatrix}.$$

This gives

$$\mathbf{S} + (-3)\mathbf{R} = \begin{pmatrix} 1 & -5 \\ 0 & 8 \end{pmatrix} + \begin{pmatrix} -9 & -21 \\ -6 & 3 \end{pmatrix}$$

$$= \begin{pmatrix} 1-9 & -5-21 \\ 0-6 & 8+3 \end{pmatrix} = \begin{pmatrix} -8 & -26 \\ -6 & 11 \end{pmatrix}.$$

(e) Writing $\mathbf{R} + 2\mathbf{R} + 2(\mathbf{R} - \mathbf{S}) = 5\mathbf{R} + (-2)\mathbf{S}$, we use our answers to parts (a) and (b):

$$5\mathbf{R} + (-2)\mathbf{S} = \begin{pmatrix} 15 & 35 \\ 10 & -5 \end{pmatrix} + \begin{pmatrix} -2 & 10 \\ 0 & -16 \end{pmatrix}$$

$$= \begin{pmatrix} 15-2 & 35+10 \\ 10+0 & -5-16 \end{pmatrix} = \begin{pmatrix} 13 & 45 \\ 10 & -21 \end{pmatrix}.$$

(f) We have

$$k\mathbf{S} = \begin{pmatrix} k\cdot 1 & k\cdot -5 \\ k\cdot 0 & k\cdot 8 \end{pmatrix} = \begin{pmatrix} k & -5k \\ 0 & 8k \end{pmatrix}.$$

5. (a) We have

$$\mathbf{A}\vec{u} = \begin{pmatrix} 2 & 5 & 7 \\ 4 & -6 & 3 \\ 16 & -5 & 0 \end{pmatrix}\begin{pmatrix} 3 \\ 2 \\ 5 \end{pmatrix}$$

$$= \begin{pmatrix} 2\cdot 3 + 5\cdot 2 + 7\cdot 5 \\ 4\cdot 3 - 6\cdot 2 + 3\cdot 5 \\ 16\cdot 3 - 5\cdot 2 + 0\cdot 5 \end{pmatrix} = \begin{pmatrix} 51 \\ 15 \\ 38 \end{pmatrix}.$$

(b) We have

$$\mathbf{B}\vec{v} = \begin{pmatrix} 8 & -6 & 0 \\ 5 & 3 & -2 \\ 3 & 7 & 12 \end{pmatrix} \begin{pmatrix} -1 \\ 0 \\ 3 \end{pmatrix}$$

$$= \begin{pmatrix} 8\cdot-1-6\cdot0+0\cdot3 \\ 5\cdot-1+3\cdot0-2\cdot3 \\ 3\cdot-1+7\cdot0+12\cdot3 \end{pmatrix} = \begin{pmatrix} -8 \\ -11 \\ 33 \end{pmatrix}.$$

(c) Letting $\vec{w} = \vec{u} + \vec{v} = (2, 2, 8)$, we have:

$$\mathbf{A}\vec{w} = \begin{pmatrix} 2 & 5 & 7 \\ 4 & -6 & 3 \\ 16 & -5 & 0 \end{pmatrix} \begin{pmatrix} 2 \\ 2 \\ 8 \end{pmatrix}$$

$$= \begin{pmatrix} 2\cdot2+5\cdot2+7\cdot8 \\ 4\cdot2-6\cdot2+3\cdot8 \\ 16\cdot2-5\cdot2+0\cdot8 \end{pmatrix} = \begin{pmatrix} 70 \\ 20 \\ 22 \end{pmatrix}.$$

Another to work this problem would be to write $\mathbf{A}(\vec{u} + \vec{v})$ as $\mathbf{A}\vec{u} + \mathbf{A}\vec{v}$ and proceed accordingly.

(d) Letting $\mathbf{C} = \mathbf{A} + \mathbf{B}$, we have

$$\mathbf{C} = \begin{pmatrix} 2 & 5 & 7 \\ 4 & -6 & 3 \\ 16 & -5 & 0 \end{pmatrix} + \begin{pmatrix} 8 & -6 & 0 \\ 5 & 3 & -2 \\ 3 & 7 & 12 \end{pmatrix} = \begin{pmatrix} 10 & -1 & 7 \\ 9 & -3 & 1 \\ 19 & 2 & 12 \end{pmatrix}.$$

We can now write $(\mathbf{A} + \mathbf{B})\vec{v}$ as $\mathbf{C}\vec{v}$, and so:

$$\mathbf{C}\vec{v} = \begin{pmatrix} 10 & -1 & 7 \\ 9 & -3 & 1 \\ 19 & 2 & 12 \end{pmatrix} \begin{pmatrix} -1 \\ 0 \\ 3 \end{pmatrix}$$

$$= \begin{pmatrix} 10\cdot-1-1\cdot0+7\cdot3 \\ 9\cdot-1-3\cdot0+1\cdot3 \\ 19\cdot-1+2\cdot0+12\cdot3 \end{pmatrix} = \begin{pmatrix} 11 \\ -6 \\ 17 \end{pmatrix}.$$

(e) From part (a) we have $\mathbf{A}\vec{u} = (51, 15, 38)$, and from part (b) we have $\mathbf{B}\vec{v} = (-8, -11, 33)$. This gives

$$\mathbf{A}\vec{u} \cdot \mathbf{B}\vec{v} = (51, 15, 38) \cdot (-8, -11, 33)$$
$$= 51\cdot-8 + 15\cdot-11 + 38\cdot33 = 681.$$

(f) We have $\vec{u} \cdot \vec{v} = 3\cdot-1 + 2\cdot0 + 5\cdot3 = 12$, and so

$$(\vec{u} \cdot \vec{v})\mathbf{A} = 12\mathbf{A} = 12\begin{pmatrix} 2 & 5 & 7 \\ 4 & -6 & 3 \\ 16 & -5 & 0 \end{pmatrix} = \begin{pmatrix} 24 & 60 & 84 \\ 48 & -72 & 36 \\ 192 & -60 & 0 \end{pmatrix}.$$

9. (a) We have

$$\begin{pmatrix} f_{\text{new}} \\ s_{\text{new}} \\ t_{\text{new}} \end{pmatrix} = \begin{pmatrix} 0.3 & 0.6 & 0.5 \\ 0.7 & 0 & 0 \\ 0 & 0.4 & 0 \end{pmatrix} \begin{pmatrix} f_{\text{old}} \\ s_{\text{old}} \\ t_{\text{old}} \end{pmatrix},$$

and so

$$f_{\text{new}} = 0.3 f_{\text{old}} + 0.6 s_{\text{old}} + 0.5 t_{\text{old}}$$
$$s_{\text{new}} = 0.7 f_{\text{old}}$$
$$t_{\text{new}} = 0.4 s_{\text{old}}.$$

From the first equation we can conclude that in a given year, 30% of the first-year insects lay eggs, as do 60% of the second-year insects and 50% of the third-year insects. From the second equation, we see that 70% of the first-year insects survive into their second year. From the third equation, we see that 40% of the second-year insects survive into their third year.

(b) We have

$$\vec{p_1} = \mathbf{T}\vec{p_0} = \begin{pmatrix} 0.3 & 0.6 & 0.5 \\ 0.7 & 0 & 0 \\ 0 & 0.4 & 0 \end{pmatrix} \begin{pmatrix} 2000 \\ 0 \\ 0 \end{pmatrix}$$

$$= \begin{pmatrix} 0.3(2000) + 0.6(0) + 0.5(0) \\ 0.7(2000) + 0(0) + 0(0) \\ 0(2000) + 0.4(0) + 0(0) \end{pmatrix} = \begin{pmatrix} 600 \\ 1400 \\ 0 \end{pmatrix}$$

$$\vec{p_2} = \mathbf{T}\vec{p_1} = \begin{pmatrix} 0.3 & 0.6 & 0.5 \\ 0.7 & 0 & 0 \\ 0 & 0.4 & 0 \end{pmatrix} \begin{pmatrix} 600 \\ 1400 \\ 0 \end{pmatrix}$$

$$= \begin{pmatrix} 0.3(600) + 0.6(1400) + 0.5(0) \\ 0.7(600) + 0(1400) + 0(0) \\ 0(600) + 0.4(1400) + 0(0) \end{pmatrix} = \begin{pmatrix} 1020 \\ 420 \\ 560 \end{pmatrix}$$

$$\vec{p_3} = \mathbf{T}\vec{p_2} = \begin{pmatrix} 0.3 & 0.6 & 0.5 \\ 0.7 & 0 & 0 \\ 0 & 0.4 & 0 \end{pmatrix} \begin{pmatrix} 1020 \\ 420 \\ 560 \end{pmatrix}$$

$$= \begin{pmatrix} 0.3(1020) + 0.6(420) + 0.5(560) \\ 0.7(1020) + 0(420) + 0(560) \\ 0(1020) + 0.4(420) + 0(560) \end{pmatrix} = \begin{pmatrix} 838 \\ 714 \\ 168 \end{pmatrix}.$$

13. (a) We have

$$\mathbf{A}\vec{v_2} = \begin{pmatrix} -2 & -1 \\ 8 & 7 \end{pmatrix} \begin{pmatrix} 1 \\ -1 \end{pmatrix} = \begin{pmatrix} -1 \\ 1 \end{pmatrix} = -1 \begin{pmatrix} -1 \\ 1 \end{pmatrix},$$

so $\lambda_2 = -1$.

(b) We have

$$\mathbf{A}\vec{v_3} = \begin{pmatrix} -2 & -1 \\ 8 & 7 \end{pmatrix} \begin{pmatrix} -3 \\ 3 \end{pmatrix} = \begin{pmatrix} 3 \\ -3 \end{pmatrix} = -1 \begin{pmatrix} -3 \\ 3 \end{pmatrix},$$

so $\lambda_3 = -1$, the same as λ_2.

The reason $\lambda_2 = \lambda_3$ is because $\vec{v_3} = -3\vec{v_2}$, that is, because $\vec{v_3}$ is a multiple of $\vec{v_2}$. Since $\mathbf{A}$ multiplies $\vec{v_2}$ by -1, it also multiples $\vec{v_3}$ by -1.

(c) Since $\vec{v}$ is an eigenvector of $\mathbf{A}$, we have $\mathbf{A}\vec{v} = \lambda\vec{v}$, $\lambda \neq 0$. We know that multiplying a vector $\vec{v}$ by a non-zero scalar λ produces a vector whose direction is the same as $\vec{v}$ and whose length is λ times the length of $\vec{v}$. This means that $\mathbf{A}\vec{v}$ is parallel to $\vec{v}$.

Solutions for Chapter 10 Review

Exercises

1. $-4\vec{i} + 8\vec{j} - 0.5\vec{i} + 0.5\vec{k} = -4.5\vec{i} + 8\vec{j} + 0.5\vec{k}$

5. $(3\vec{j} - 2\vec{k} + \vec{i}) \cdot (4\vec{k} - 2\vec{i} + 3\vec{j}) = (\vec{i} + 2\vec{j} - 3\vec{k}) \cdot (-2\vec{i} + 3\vec{j} + 4\vec{k}) = -2 + 6 - 12 = -8.$

9. $\vec{a} + \vec{b} = (5, 1, 0) + (2, -1, 9) = (7, 0, 9)$

13. $2\vec{a} - 3(\vec{b} - \vec{c}) = 2(5, 1, 0) - 3((2, -1, 9) - (1, 1, 2)) = (10, 2, 0) - 3(1, -2, 7) = (7, 8, -21).$

17. $\vec{a} = \vec{b} = \vec{c} = 3\vec{k},\quad \vec{d} = 2\vec{i} + 3\vec{k},\quad \vec{e} = \vec{j},\quad \vec{f} = -2\vec{i}.$

Problems

21. (a) We have

$$\vec{E}_{\max} \cdot \vec{T} = (40, 40, 30, 15, 10) \cdot (30, 30, 40, 80, 120)$$
$$= 40 \cdot 30 + 40 \cdot 30 + 30 \cdot 40 + 15 \cdot 80 + 10 \cdot 120 = 6000,$$

so the total weekly tuition for all students is $6000.

(b) We have

$$\vec{T}_{\text{new}} = (30, 30, 40, 80, 120) + (5, 5, 10, 20, 30) = (35, 35, 50, 100, 150).$$

This is the new tuition vector. We see that

$$\vec{E}_{\max} \cdot \vec{R} = (40, 40, 30, 15, 10) \cdot (5, 5, 10, 20, 30)$$
$$= 40 \cdot 5 + 40 \cdot 5 + 30 \cdot 10 + 15 \cdot 20 + 10 \cdot 30 = 1300,$$

which tells us that the total change in weekly tuition is $1300. Also,

$$\vec{E}_{\max} \cdot \vec{T}_{\text{new}} = (40, 40, 30, 15, 10) \cdot (35, 35, 50, 100, 150)$$
$$= 40 \cdot 35 + 40 \cdot 35 + 30 \cdot 50 + 15 \cdot 100 + 10 \cdot 150 = 7300,$$

which is the new tuition.

Notice that $\vec{E}_{\max} \cdot (\vec{T} + \vec{R}) = \vec{E}_{\max} \cdot \vec{T} + \vec{E}_{\max} \cdot \vec{R}$, as required by the distributive law for the dot product.

25. The velocity vector of the plane with respect to the calm air has the form

$$\vec{v} = a\vec{i} + 80\vec{k} \text{ where } \|\vec{v}\| = 480.$$

(See Figure 10.13.) Therefore $\sqrt{a^2 + 80^2} = 480$ so $a = \sqrt{480^2 - 80^2} \approx 473.286$ km/hr. We conclude that $\vec{v} \approx 473.286\vec{i} + 80\vec{k}$.

The wind vector is

$$\vec{w} = 100(\cos 45°)\vec{i} + 100(\sin 45°)\vec{j}$$
$$\approx 70.711\vec{i} + 70.711\vec{j}.$$

The velocity vector of the plane with respect to the ground is then

$$\vec{v} + \vec{w} = (473.286\vec{i} + 80\vec{k}) + (70.711\vec{i} + 70.711\vec{j})$$
$$= 544\vec{i} + 70.711\vec{j} + 80\vec{k}.$$

From Figure 10.14, we see that the velocity relative to the ground is

$$543.997\vec{i} + 70.7\vec{j}.$$

The ground speed is therefore $\sqrt{543.997^2 + 70.7^2} \approx 548.573$ km/hr.

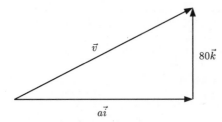

Figure 10.13: Side view

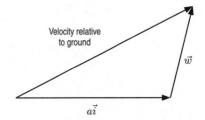

Figure 10.14: Top view

29. See Figure 10.15, where $\vec{g}$ is the acceleration due to gravity, and $g = \|\vec{g}\|$.
    If $\theta = 0$ (the plank is at ground level), the sliding force is $F = 0$.
    If $\theta = \pi/2$ (the plank is vertical), the sliding force equals g, the force due to gravity.
    Therefore, we can guess that F is proportional with $\sin\theta$:

$$F = g\sin\theta.$$

This agrees with the bounds at $\theta = 0$ and $\theta = \pi/2$, and with the fact that the sliding force is smaller than g between 0 and $\pi/2$.

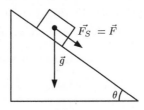

Figure 10.15

33. Suppose θ is the angle between $\vec{u}$ and $\vec{v}$.

    (a) By the definition of scalar multiplication, we know that $-\vec{v}$ is in the opposite direction of $\vec{v}$, so the angle between $\vec{u}$ and $-\vec{v}$ is $\pi - \theta$. (See Figure 10.16.) Hence,

$$\vec{u} \cdot (-\vec{v}) = \|\vec{u}\|\|-\vec{v}\|\cos(\pi - \theta)$$
$$= \|\vec{u}\|\|\vec{v}\|(-\cos\theta)$$
$$= -(\vec{u} \cdot \vec{v}).$$

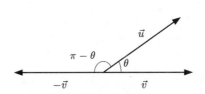

Figure 10.16

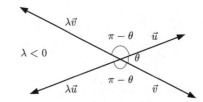

Figure 10.17

(b) If $\lambda < 0$, the angle between $\vec{u}$ and $\lambda\vec{v}$ is $\pi - \theta$, and so is the angle between $\lambda\vec{u}$ and $\vec{v}$. (See Figure 10.17.) So we have,

$$\begin{aligned}
\vec{u} \cdot (\lambda\vec{v}) &= \|\vec{u}\|\|\lambda\vec{v}\|\cos(\pi - \theta) \\
&= |\lambda|\|\vec{u}\|\|\vec{v}\|(-\cos\theta) \\
&= -\lambda\|\vec{u}\|\|\vec{v}\|(-\cos\theta) \quad \text{since } |\lambda| = -\lambda \\
&= \lambda\|\vec{u}\|\|\vec{v}\|\cos\theta \\
&= \lambda(\vec{u} \cdot \vec{v}).
\end{aligned}$$

By a similar argument, we have

$$\begin{aligned}
(\lambda\vec{u}) \cdot \vec{v} &= \|\lambda\vec{u}\|\|\vec{v}\|\cos(\pi - \theta) \\
&= -\lambda\|\vec{u}\|\|\vec{v}\|(-\cos\theta) \\
&= \lambda(\vec{u} \cdot \vec{v}).
\end{aligned}$$

37. Using the result of Problem 36, we have $\overrightarrow{AC} = \vec{w} + \vec{n} - \vec{m} = 3\vec{n} - 3\vec{m}$; $\overrightarrow{AB} = \vec{v} + \vec{m} + \vec{n} = 3\vec{m} + \vec{n}$; $\overrightarrow{AD} = \vec{v} + \vec{m} - (\vec{n} - \vec{m}) = 4\vec{m} - \vec{n}$; $\overrightarrow{BD} = (-\vec{n}) - (\vec{n} - \vec{m}) = \vec{m} - 2\vec{n}$.

CHECK YOUR UNDERSTANDING

1. False. The length $\|0.5\vec{i} + 0.5\vec{j}\| = \sqrt{(0.5)^2 + (0.5)^2} = \sqrt{0.5} \neq 1$.

5. False. The dot product of two vectors is a scalar.

9. False. The length of $\vec{v}$ is the square root of $\vec{v} \cdot \vec{v}$.

13. False. Multiplication is not commutative. For example, if $A = \begin{pmatrix} 0 & 1 \\ 1 & 0 \end{pmatrix}$ and $B = \begin{pmatrix} 2 & 0 \\ 0 & 3 \end{pmatrix}$, then $AB = \begin{pmatrix} 0 & 3 \\ 2 & 0 \end{pmatrix}$ and $BA = \begin{pmatrix} 0 & 2 \\ 3 & 0 \end{pmatrix}$.

CHAPTER ELEVEN

Solutions for Section 11.1

Exercises

1. Not arithmetic. The differences are 5, 4, 3.

5. Geometric. The ratios of successive terms are all 3.

9. Arithmetic, with $a = 6$, $d = 3$, so $a_n = 6 + (n-1)3 = 3 + 3n$.

13. Geometric, since the ratios of successive terms are all $1/2$. Thus, $a = 4$ and $r = 1/2$, so $a_n = 4(\frac{1}{2})^{n-1}$.

17. Geometric, since the ratios of successive terms are all $1/1.2$. Thus, $a = 1$ and $r = 1/1.2$, so $a_n = 1(1/1.2)^{n-1} = 1/(1.2)^{n-1}$.

Problems

21. Arithmetic, because points lie on a line. The sequence is decreasing, so $d < 0$.

25. Since $a_1 = 3$ and $a_n = 2a_{n-1}$, we have $a_2 = 2a_1 = 6$, $a_3 = 2a_2 = 12$, $a_4 = 2a_3 = 24$.

   If we do not multiply out, we can see the general pattern more easily:

$$a_2 = 2 \cdot 3, \quad a_3 = 2 \cdot 2 \cdot 3 = 2^2 \cdot 3, \quad a_4 = 2 \cdot 2^2 \cdot 3 = 2^3 \cdot 3.$$

   Thus a_n is a geometric sequence, $a_n = 2^{n-1} \cdot 3$.

29. (a) Since there are more people with incomes less than or equal to \$50,000 than there are people with incomes less than or equal to \$40,000 (because all those with incomes less than or equal to \$40,000 also have incomes less than or equal to \$50,000), we have $a_{40} < a_{50}$.

(b) The quantity $a_{50} - a_{40}$ represents the fraction of the US population who have incomes between \$40,000 and \$50,000.

(c) No, because there are some people (for example, children) whose income is zero. This fraction of the population is included in a_n for all n. Thus, a_n is never 0.

(d) The value of a_n increases as n increases. If n is larger than all incomes in the US, then $a_n = 1$.

Solutions for Section 11.2

Exercises

1. One way to work this problem is to complete the first few values of a_n, and then continue the pattern across the row. Then, the values of S_n can be found by summing the values of a_n from left to right. See Table 11.1. We see that $a_1 = 2$ and that $d = a_2 - a_1 = 7 - 2 = 5$.

Table 11.1

n	1	2	3	4	5	6	7	8
a_n	2	7	12	17	22	27	32	37
S_n	2	9	21	38	60	87	119	156

5. One way to work this problem is to complete the first few values of a_n, and then continue the pattern across the row. Then, the values of S_n can be found by summing the values of a_n from left to right. See Table 11.2. We knew that $S_6 = 201$, $S_7 = 273$, and $S_8 = 356$. Note that $S_7 = S_6 + a_7$, and so $273 = 201 + a_7$, which means that $a_7 = 72$. Similarly, we know that $S_8 = S_7 + a_8$, and so $356 = 273 + a_8$, which means that $a_8 = 83$. Now we know two consecutive values of a_n, and so we can find d: we have $d = a_8 - a_7 = 83 - 72 = 11$. Finally, we can find a_1 as follows:

$$a_7 = a_1 + (7 - 1)d$$
$$72 = a_1 + 6 \cdot 11$$
$$a_1 = 6.$$

Table 11.2

n	1	2	3	4	5	6	7	8
a_n	6	17	28	39	50	61	72	83
S_n	6	23	51	90	140	201	273	356

9. $\displaystyle\sum_{n=1}^{7} (-1)^{n-1} 2^n = (-1)^0 2^1 + (-1)^1 2^2 + (-1)^2 2^3 + \cdots (-1)^6 2^7.$

13. The first term is 30 and then each term is five less than the preceding term until we reach $5 = 30 - 5 \cdot 5$. So, a possible solution is

$$\sum_{n=0}^{5} (30 - 5n).$$

17. $\sum_{i=1}^{30} (5i + 10) = 15 + 20 + 25 + \cdots$. We use the formula to find the sum of this arithmetic series

$$S_{30} = \frac{1}{2} \cdot 30(2 \cdot 15 + 29 \cdot 5) = 2625.$$

Problems

21. (a) **(i)** From the table, we have $S_4 = 226.6$, the population of the US in millions 4 decades after 1940, that is, in 1980. Similarly, $S_5 = 248.7$, the population in millions in 1990, and $S_6 = 281.4$, the population in 2000.

(ii) We have $a_2 = S_2 - S_1 = 179.3 - 150.7 = 28.6$; that is, the increase in the US population in millions in the 1950s.
Similarly, $a_5 = S_5 - S_4 = 248.7 - 226.6 = 22.1$, the population increase in millions during the 1980s. In the same way, $a_6 = S_6 - S_5 = 281.4 - 248.7 = 32.7$, the population increase during the 1990s.

(iii) Using the answer to (ii), we have $a_6/10 = 32.7/10 = 3.27$, the average yearly population growth during the 1990s.

(iv) We have
S_n = US population, in millions, n decades after 1940.
$a_n = S_n - S_{n-1}$ = growth in US population in millions, during the nth decade after 1940.
$a_n/10$ = Average yearly growth, in millions, during the nth decade after 1940.

25. We know that $S_7 = 16 + 48 + 80 + \cdots + 208$. Here, $a_1 = 16$, $n = 7$, and $d = 32$. This gives

$$S_7 = \frac{1}{2} n (2a_1 + (n - 1)d) = \frac{1}{2} \cdot 7 (2 \cdot 16 + (7 - 1)32) = 784.$$

29. (a) On the nth round, the boy gives his sister 1 M&M and takes n for himself.

(b) After n rounds, the sister has n M&Ms. The number of M&Ms that the boy has is given by the arithmetic series

$$\text{Number of M\&Ms} = 1 + 2 + 3 + \cdots + n = \sum_{i=1}^{n} i = \frac{n(n + 1)}{2}.$$

Solutions for Section 11.3

Exercises

1. Yes, $a = 2$, ratio $= 1/2$.

5. $3 + \dfrac{3}{2} + \dfrac{3}{4} + \dfrac{3}{8} \cdots + \dfrac{3}{2^{10}} = 3\left(1 + \dfrac{1}{2} + \cdots + \dfrac{1}{2^{10}}\right) = \dfrac{3\left(1 - \frac{1}{2^{11}}\right)}{1 - \frac{1}{2}} = 5.997.$

9. Note that we are seeking the sum of the first eight terms. Thus,

$$S_8 = \frac{2\left(1 - \left(\frac{3}{4}\right)^8\right)}{1 - \frac{3}{4}} = 7.199.$$

13. We again need powers of -1 to create the alternating pattern. The first term is 32, and each term is $1/2$ of the preceding term, so we could use powers of $1/2$ to create the series. One possible answer is: $\displaystyle\sum_{n=0}^{5} (-1)^n 32 \left(\frac{1}{2}\right)^n.$

Problems

17. Let B_n be the balance in dollars right after the nth deposit. Then

$$B_1 = 1000$$
$$B_2 = 1000(1.03) + 1000$$
$$B_3 = 1000(1.03)^2 + 1000(1.03) + 1000$$
$$\vdots$$
$$B_{20} = 1000((1.03)^{19} + (1.03)^{18} + \cdots + 1).$$

Using the formula for the sum of a finite geometric series,

$$B_{20} = \frac{1000(1 - (1.03)^{20})}{1 - 1.03} = 26{,}870.37 \text{ dollars.}$$

Solutions for Section 11.4

Exercises

1. Yes, $a = 1$, ratio $= -x$.

5. Sum $= \dfrac{1}{1 - (-x)} = \dfrac{1}{1 + x}$, $|x| < 1$.

9. $\displaystyle\sum_{i=0}^{\infty} \frac{3^i + 5}{4^i} = \sum_{i=0}^{\infty} \left(\frac{3}{4}\right)^i + \sum_{i=0}^{\infty} \frac{5}{4^i},$ a sum of two geometric series.

$$\sum_{i=0}^{\infty} \left(\frac{3}{4}\right)^i = \frac{1}{1 - \frac{3}{4}} = 4,$$

$$\sum_{i=0}^{\infty} \frac{5}{4^i} = \frac{5}{1 - \frac{1}{4}} = \frac{20}{3},$$

so $\displaystyle\sum_{i=0}^{\infty} \frac{3^i + 5}{4^i} = 4 + \frac{20}{3} = \frac{32}{3}.$

Problems

13. $0.122222\ldots = 0.1 + \dfrac{2}{100} + \dfrac{2}{1000} + \dfrac{2}{10000} + \dfrac{2}{100000} + \cdots$. Thus,

$$S = 0.1 + \frac{\frac{2}{100}}{1 - \frac{1}{10}} = \frac{1}{10} + \frac{2}{90} = \frac{11}{90}.$$

17. (a)

$$P_1 = 0$$
$$P_2 = 250(0.04)$$
$$P_3 = 250(0.04) + 250(0.04)^2$$
$$P_4 = 250(0.04) + 250(0.04)^2 + 250(0.04)^3$$
$$\vdots$$
$$P_n = 250(0.04) + 250(0.04)^2 + \cdots + 250(0.04)^{n-1}$$

(b) Factoring our formula for P_n, we see that it involves a geometric series of $n-2$ terms:

$$P_n = 250(0.04) \underbrace{\left[1 + 0.04 + (0.04)^2 + \cdots + (0.04)^{n-2} \right]}_{n-2 \text{ terms}}.$$

The sum of this series is given by

$$1 + 0.04 + (0.04)^2 + \cdots + (0.04)^{n-2} = \frac{1 - (0.04)^{n-1}}{1 - 0.04}.$$

Thus,

$$P_n = 250(0.04) \left(\frac{1 - (0.04)^{n-1}}{1 - 0.04} \right)$$
$$= 10 \left(\frac{1 - (0.04)^{n-1}}{1 - 0.04} \right).$$

(c) In the long run, that is, as $n \to \infty$, we know that $(0.04)^{n-1} \to 0$, and so

$$P_n = 10 \left(\frac{1 - (0.04)^{n-1}}{1 - 0.04} \right) \to 10 \left(\frac{1 - 0}{1 - 0.04} \right) = 10.417.$$

Thus, P_n gets closer to 10.417 and Q_n gets closer to 260.42. We'd expect these limits to differ because one is right before taking a tablet and one is right after. We'd expect the difference between them to be exactly 250 mg, the amount of ampicillin in one tablet.

21.

$$\text{Present value of first coupon} = \frac{50}{1.04}$$
$$\text{Present value of second coupon} = \frac{50}{(1.04)^2}, \text{ etc.}$$

$$\text{Total present value} = \underbrace{\frac{50}{1.04} + \frac{50}{(1.04)^2} + \cdots + \frac{50}{(1.04)^{10}}}_{\text{coupons}} + \underbrace{\frac{1000}{(1.04)^{10}}}_{\text{principal}}$$

$$= \frac{50}{1.04} \left(1 + \frac{1}{1.04} + \cdots + \frac{1}{(1.04)^9} \right) + \frac{1000}{(1.04)^{10}}$$

$$= \frac{50}{1.04} \left(\frac{1 - \left(\frac{1}{1.04} \right)^{10}}{1 - \frac{1}{1.04}} \right) + \frac{1000}{(1.04)^{10}}$$

$$= 405.545 + 675.564$$

$$= \$1081.11.$$

Solutions for Chapter 11 Review——————————————————————

Exercises

1. We start with 100, decrease by tens until we reach 0, which is $100 - 10(10)$. A possible answer is

$$\sum_{n=0}^{10}(100 - 10n).$$

5. Yes, $a = 1$, ratio $= 2z$.

Problems

9. (a) The sequence is 1, 4, 9, 16. The nth grid has n dots per side, that is, n rows of n dots each, for a total of n^2 dots.

    (b) The number of black dots in each grid is 1, 3, 5, 7, ... This appears to be the sequence of odd numbers. To see that this is true, notice (in Figure 11.1) that starting with a grid with 2 dots on each side, we add 2 dots to the right, 2 dots to the top, and 1 dot to the corner. Likewise, starting with a grid with 3 dots on each side, we add 3 dots to the right, 3 dots to the top, and 1 dot to the corner:

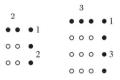

Figure 11.1

        In general, starting with a grid with $n - 1$ dots on each side, we add $n - 1$ dots to the right, $n - 1$ dots to the top, and 1 dot to the corner to obtain a grid with n dots on each side:

$$\text{Number of black dots in } n^{\text{th}} \text{ grid} = (n - 1) + (n - 1) + 1 = 2n - 1.$$

The nth grid has $2n - 1$ black dots.

    (c) Notice that the total number of dots in the second grid equals the total number of *black* dots in the first two grids. Likewise, the total number of dots in the third grid equals the total number of *black* dots in the first three grids, and so on. Letting a_n be the number of black dots in the nth grid, and S_n be the total number of dots in the nth grid, we see that

$$S_n = a_1 + a_2 + \cdots + a_n.$$

From part (b), we have $a_n = 2n - 1$. If we rewrite this as $a_n = 2(n - 1) + 1$, we have $a_n = a_1 + (n - 1)d$ where $a_1 = 1$ and $d = 2$. Using our formula for the sum of an arithmetic series, this gives

$$\begin{aligned}
S_n &= \frac{1}{2}n(2a_1 + (n - 1)d) \\
&= \frac{1}{2}n(2 + (n - 1)(2)) \\
&= n + n(n - 1) \\
&= n^2,
\end{aligned}$$

which agrees with our formula in part (a), thus verifying the relationship.

13. (a) Since $100,000 earns $0.03 \cdot 100,000 = \$3000$ interest in one year, and $98,000 earns only slightly less ($2940), the withdrawal is smaller than the interest earned, so we expect the balance to increase with time. Thus, we expect the balance to be higher after the second withdrawal.

(b) Let B_n be the balance in the account in dollars right after the nth withdrawal of $2000. Then

$$B_1 = 100{,}000 - 2000 = 98{,}000$$
$$B_2 = B_1(1.03) - 2000 = 98{,}000(1.03) - 2000$$
$$B_3 = B_2(1.03) - 2000 = 98{,}000(1.03)^2 - 2000(1.03) - 2000$$
$$B_4 = B_3(1.03) - 2000 = 98{,}000(1.03)^3 - 2000((1.03)^2 + 1.03 + 1)$$
$$B_5 = B_4(1.03) - 2000 = 98{,}000(1.03)^4 - 2000((1.03)^3 + (1.03)^2 + 1.03 + 1)$$
$$\vdots$$
$$B_{20} = 98{,}000(1.03)^{19} - 2000((1.03)^{18} + (1.03)^{17} + \cdots + 1.03 + 1).$$

Using the formula for the sum of a finite geometric series, we have

$$B_{20} = 98{,}000(1.03)^{19} - \frac{2000(1 - (1.03)^{19})}{1 - 1.03} = 121{,}609.86 \text{ dollars.}$$

(c) With withdrawals of $2000 a year, the balance increases. The balance remains constant if the withdrawals exactly balance the interest earned; larger withdrawals cause the balance to decrease. If the largest withdrawal is $\$x$, then $B_1 = 100{,}000 - x$. The interest earned on this equals x, so

$$0.03(100{,}000 - x) = x.$$

Thus

$$0.03 \cdot 100{,}000 - 0.03x = x$$
$$x = \frac{0.03 \cdot 100{,}000}{1.03} = 2912.62 \text{ dollars.}$$

17. (a)

$$\text{Total amount of money deposited} = 100 + 92 + 84.64 + \cdots$$
$$= 100 + 100(0.92) + 100(0.92)^2 + \cdots$$
$$= \frac{100}{1 - 0.92} = 1250 \quad \text{dollars.}$$

(b) Credit multiplier $= 1250/100 = 12.50$

The 12.50 is the factor by which the bank has increased its deposits, from $100 to $1250.

CHECK YOUR UNDERSTANDING

1. True. $a_1 = (1)^2 + 1 = 2$.

5. True. The differences between successive terms are all 1.

9. True. The first partial sum is just the first term of the sequence.

13. True. The sum is n terms of 3. That is, $3 + 3 + \cdots + 3 = 3n$.

17. False. The terms of the series can be negative so partial sums can decrease.

21. True. If $a = 1$ and $r = -\frac{1}{2}$, it can be written $\sum_{i=0}^{5}(-\frac{1}{2})^i$.

25. False. If payments are made at the end of each year, after 20 years, the balance at 5% is about $66,000, while at 10% it would be about $115,000. If payments are made at the start of each year, the corresponding figures are $69,000 and $126,000.

29. False. The series does not converge since the odd terms (Q_1, Q_3, etc.) are all -1.

33. False. An arithmetic series with $d \neq 0$ diverges.

CHAPTER TWELVE

Solutions for Section 12.1

Exercises

1. The graph of the parametric equations is in Figure 12.1.

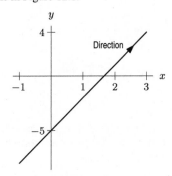

Figure 12.1

Since $x = t + 1$, we have $t = x - 1$. Substitute this into the second equation:

$$y = 3t - 2$$
$$y = 3(x - 1) - 2$$
$$y = 3x - 5.$$

5. The graph of the parametric equations is in Figure 12.2.

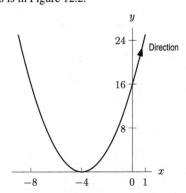

Figure 12.2

Since $x = t - 3$, we have $t = x + 3$. Substitute this into the second equation:

$$y = t^2 + 2t + 1$$
$$y = (x + 3)^2 + 2(x + 3) + 1$$
$$= x^2 + 8x + 16 = (x + 4)^2.$$

9. The graph of the parametric equations is in Figure 12.3.

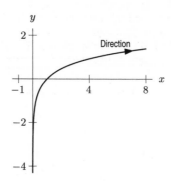

Figure 12.3

Since $x = t^3$, we take the natural log of both sides and get $\ln x = 3 \ln t$ or $\ln t = 1/3 \ln x$. We are given that $y = 2 \ln t$, thus,

$$y = 2 \left(\frac{1}{3} \ln x \right) = \frac{2}{3} \ln x.$$

13. Between times $t = 0$ and $t = 1$, x goes at a constant rate from 0 to 1 and y goes at a constant rate from 1 to 0. So the particle moves in a straight line from $(0, 1)$ to $(1, 0)$. Similarly, between times $t = 1$ and $t = 2$, it goes in a straight line to $(0, -1)$, then to $(-1, 0)$, then back to $(0, 1)$. So it traces out the diamond shown in Figure 12.4.

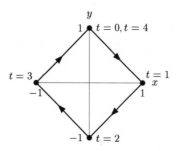

Figure 12.4

Problems

17. (a) We can replace x with t and $t + 1$ to get two parameterizations:

$$x = t, \quad y = t^2 \qquad \text{and} \qquad x = t + 1, \quad y = (t + 1)^2.$$

Alternatively, the second parameterization could look totally different:

$$x = t^3, \quad y = t^6.$$

(b) Altering the answers to part (a) gives:

$$x = t, \quad y = (t + 2)^2 + 1 \qquad \text{and} \qquad x = t + 1, \quad y = (t + 3)^2 + 1.$$

21. Let $f(t) = \ln t$. The particle is moving counterclockwise when $t > 0$. Any other time, when $t \le 0$, the position is not defined.

25. (a) Since the x-coordinate and the y-coordinate are always the same (they both equal t), the bug follows the path $y = x$.

(b) The bug starts at $(1, 0)$ because $\cos 0 = 1$ and $\sin 0 = 0$. Since the x-coordinate is $\cos x$, and the y-coordinate is $\sin x$, the bug follows the path of a unit circle, traveling counterclockwise. It reaches the starting point of $(1, 0)$ when $t = 2\pi$, because $\sin t$ and $\cos t$ are periodic with period 2π.

(c) Now the x-coordinate varies from 1 to -1, while the y-coordinate varies from 2 to -2; otherwise, this is much like part (b) above. If we plot several points, the path looks like an ellipse, which is a circle stretched out in one direction.

29. For $0 \le t \le 2\pi$, the graph is in Figure 12.5.

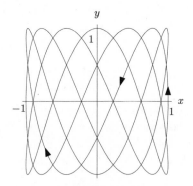

Figure 12.5

33. The particle moves back and forth between -1 and 1. See Figure 12.6.

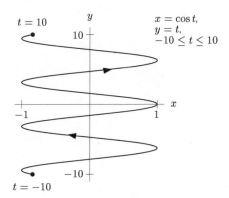

Figure 12.6

Solutions for Section 12.2

Exercises

1. Dividing by 4 and rewriting the equation as

$$x^2 + y^2 = \frac{9}{4},$$

we see that the center is $(0, 0)$, and the radius is $3/2$.

5. We consider $x = 4\cos t, y = 4\sin t$ for $0 \le t \le 2\pi$. However, since y increases from 0 as t increases from 0, this goes counterclockwise. We can use

$$x = 4\cos t, \quad y = -4\sin t \quad \text{for } 0 \le t \le 2\pi.$$

9. We can use $x = 3 + 5\cos t, y = 4 + 5\sin t$ for $0 \le t \le 2\pi$.

Problems

13. **(a)** Center is $(2, -4)$ and radius is $\sqrt{20}$.
    (b) Rewriting the original equation and completing the square, we have

$$2x^2 + 2y^2 + 4x - 8y = 12$$
$$x^2 + y^2 + 2x - 4y = 6$$
$$(x^2 + 2x + 1) + (y^2 - 4y + 4) - 5 = 6$$
$$(x + 1)^2 + (y - 2)^2 = 11.$$

So the center is $(-1, 2)$, and the radius is $\sqrt{11}$.

17. Since $y - 3 = \sin t$ and $x = 4\sin^2 t$, this parameterization traces out the parabola $x = 4(y - 3)^2$ for $2 \le y \le 4$.

21. Implicit: $x^2 - 2x + y^2 = 0, y < 0$. Explicit: $y = -\sqrt{-x^2 + 2x}, 0 \le x \le 2$. Parametric: The curve is the lower half of a circle centered at $(1, 0)$ with radius 1, so $x = 1 + \cos t, y = \sin t$, for $\pi \le t \le 2\pi$.

Solutions for Section 12.3

Exercises

1. **(a)** The center is at the origin. The diameter in the x-direction is 24 and the diameter in the y-direction is 10.
   (b) The equation of the ellipse is

$$\frac{x^2}{12^2} + \frac{y^2}{5^2} = 1 \quad \text{or} \quad \frac{x^2}{144} + \frac{y^2}{25} = 1.$$

5. We can use $x = 12\cos t, y = 5\sin t$ for $0 \le t \le 2\pi$.

9. This ellipse has a graph which is the same curve in the xy plane, but it is traced out at twice the speed. At $t = \pi$, this parameterization has already returned to its starting point $(9, 4)$. This parameterization traces out the ellipse twice during $0 \le t \le 2\pi$.

Problems

13. Factoring out the 4 from $4x^2 + 16 = 4(x^2 + 4x)$ and completing the square on $x^2 + 4x$ and $y^2 + 2y$:

$$4(x^2 + 4x) + y^2 + 2y + 13 = 0$$
$$4((x + 2)^2 - 4) + (y + 1)^2 - 1 + 13 = 0$$
$$4(x + 2)^2 - 16 + (y + 1)^2 - 1 + 13 = 0$$
$$4(x + 2)^2 + (y + 1)^2 - 4 = 0.$$

Moving the -4 to the right and dividing by 4 to get 1 on the right side:

$$4(x + 2)^2 + (y + 1)^2 = 4$$
$$(x + 2)^2 + \frac{(y + 1)^2}{4} = 1.$$

The center is $(-2, -1)$, and $a = 1, b = 2$.

17. (a) The center is $(-1, 3)$, major axis $a = \sqrt{6}$, minor axis $b = 2$. Figure 12.7 shows the graph.

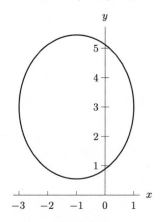

Figure 12.7

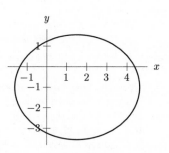

Figure 12.8

(b) Rewriting the equation and completing the square, we have

$$2x^2 + 3y^2 - 6x + 6y = 12$$
$$2(x^2 - 3x) + 3(y^2 + 2y) = 12$$
$$2(x^2 - 3x + \frac{9}{4}) + 3(y^2 + 2y + 1) - \frac{9}{2} - 3 = 12$$
$$2(x - \frac{3}{2})^2 + 3(y + 1)^2 = \frac{39}{2}$$
$$\frac{(x - \frac{3}{2})^2}{(39/4)} + \frac{(y + 1)^2}{(13/2)} = 1.$$

So the center is $(3/2, -1)$, major axis is $a = \sqrt{39}/2$, minor axis is $b = \sqrt{13/2}$. Figure 12.8 shows the graph.

21. We have

$$Ax^2 - Bx + y^2 = r_0^2$$
$$A\left(x^2 - \frac{B}{A}x\right) + y^2 = r_0^2$$
$$A\left(x^2 - \frac{B}{A}x + \left(\frac{B}{2A}\right)^2\right) + y^2 = r_0^2 + \frac{B^2}{4A} \qquad \text{completing the square}$$
$$A\left(x - \frac{B}{2A}\right)^2 + y^2 = \frac{4Ar_0^2 + B^2}{4A}.$$

Dividing both sides by $(4Ar_0^2 + B^2)/(4A)$, we obtain

$$\frac{(x - B/2A)^2}{a^2} + \frac{y^2}{b^2} = 1,$$

where $a^2 = (4Ar_0^2 + B^2)/(4A^2)$ and $b^2 = (4Ar_0^2 + B^2)/(4A)$.

Solutions for Section 12.4

Exercises

1. (a) The vertices are at $(5, 0)$ and $(-5, 0)$. The center is at the origin.

(b) The asymptotes have slopes $1/5$ and $-1/5$. The equations of the asymptotes are

$$y = \frac{1}{5}x \quad \text{and} \quad y = -\frac{1}{5}x.$$

(c) The equation of the hyperbola is

$$\frac{x^2}{5^2} - \frac{y^2}{1^2} = 1 \quad \text{or} \quad \frac{x^2}{25} - y^2 = 1.$$

Problems

5. The hyperbola is centered at the origin, and $a = 5, b = 1$. We can use $x = 5 \sec t = 5/\cos t, y = \tan t$.

If $0 < t < \pi/2$, then $x > 0, y > 0$, so we have Quadrant I.
If $\pi/2 < t < \pi$, then $x < 0, y < 0$, so we have Quadrant III.
If $\pi < t < 3\pi/2$, then $x < 0, y > 0$, so we have Quadrant II.
If $3\pi/2 < t < 2\pi$, then $x > 0, y < 0$, so we have Quadrant IV.
So the right half is given by $0 \le t < \pi/2$ together with $3\pi/2 < t < 2\pi$.

9. Factoring out -1 from $-y^2 + 4y = -(y^2 - 4y)$ and completing the square on $x^2 - 2x$ and $y^2 - 4y$ gives

$$\frac{1}{4}(x^2 - 2x) - (y^2 - 4y) = \frac{19}{4}$$

$$\frac{1}{4}((x - 1)^2 - 1) - ((y - 2)^2 - 4) = \frac{19}{4}$$

$$\frac{(x - 1)^2}{4} - \frac{1}{4} - (y - 2)^2 + 4 = \frac{19}{4}$$

$$\frac{(x - 1)^2}{4} - (y - 2)^2 = 1.$$

The center is $(1, 2)$, the hyperbola opens right-left, and $a = 2, b = 1$.

13. Factoring out 4 from $4x^2 - 8x = 4(x^2 - 2x)$ and 36 from $36y^2 - 36y = 36(y^2 - y)$ and completing the square on $x^2 - 2x$ and $y^2 - y$ gives

$$4(x^2 - 2x) = 36(y^2 - y) - 31$$

$$4((x - 1)^2 - 1) = 36\left(\left(y - \frac{1}{2}\right)^2 - \frac{1}{4}\right) - 31$$

$$4(x - 1)^2 - 4 = 36\left(y - \frac{1}{2}\right)^2 - 9 - 31$$

$$4(x - 1)^2 = 36\left(y - \frac{1}{2}\right)^2 - 36.$$

Moving $36(y - \frac{1}{2})^2$ to the left and dividing by -36 to get 1 on the right:

$$\frac{4(x - 1)^2}{-36} - \frac{36\left(y - \frac{1}{2}\right)^2}{-36} = -\frac{36}{-36}$$

$$-\frac{(x - 1)^2}{9} + \left(y - \frac{1}{2}\right)^2 = 1$$

$$\left(y - \frac{1}{2}\right)^2 - \frac{(x - 1)^2}{9} = 1.$$

The center is $(1, \frac{1}{2})$, the hyperbola opens up-down, and $a = 3, b = 1$.

17. (a) The center is $(-5, 2)$, vertices are $(-5 + \sqrt{6}, 2)$ and $(-5 - \sqrt{6}, 2)$. The asymptotes are $y = \pm\frac{2}{\sqrt{6}}(x + 5) + 2$.
Figure 12.9 shows the hyperbola.

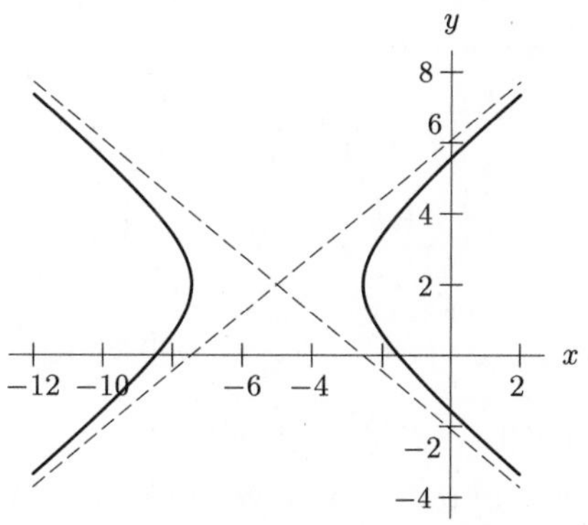

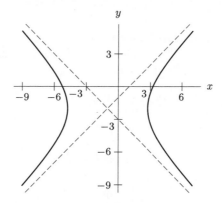

Figure 12.9

(b) Rewriting the equation and competing the square, we have

$$x^2 - y^2 + 2x = 4y + 17$$
$$x^2 - y^2 + 2x - 4y = 17$$
$$(x^2 + 2x + 1) - (y^2 + 4y + 4) + 3 = 17$$
$$(x + 1)^2 - (y + 2)^2 = 14$$
$$\frac{(x + 1)^2}{14} - \frac{(y + 2)^2}{14} = 1.$$

Thus the center is $(-1, -2)$; vertices are $(-1 - \sqrt{14}, -2)$ and $(-1 + \sqrt{14}, -2)$; asymptotes are $y = \pm(x + 1) - 2$, that is, $y = x - 1$ and $y = -x - 3$. Figure 12.10 shows the hyperbola.

Figure 12.10

Solutions for Section 12.5

Exercises

1. Substitute $x = 0$ into the formula for $\sinh x$. This yields

$$\sinh 0 = \frac{e^0 - e^{-0}}{2} = \frac{1 - 1}{2} = 0.$$

5. We use $x = \sinh t$, $y = \cosh t$, for $-\infty < t < \infty$.

9. We use $x = 1 + 2\sinh t$, $y = -1 - 3\cosh t$, for $-\infty < t < \infty$.

13. Divide by 6 to rewrite the equation as

$$\frac{12(x-1)^2}{6} - \frac{3(y+2)^2}{6} = 1, \quad x > 1$$

$$\frac{(x-1)^2}{1/2} - \frac{(y+2)^2}{2} = 1, \quad x > 1,$$

so use $x = 1 + (1/\sqrt{2})\cosh t$, $y = -2 + \sqrt{2}\sinh t$, for $-\infty < t < \infty$.

Problems

17. Factoring out 2 from $2x^2 - 12x = 2(x^2 - 6x)$ and 4 from $4y^2 + 4y = 4(y^2 + y)$ and completing the square on $x^2 - 6x$ and $y^2 + y$ gives

$$25 + 2(x^2 - 6x) = 4(y^2 + y), \quad y > -\frac{1}{2}$$

$$25 + 2((x-3)^2 - 9) = 4\left(\left(y + \frac{1}{2}\right)^2 - \frac{1}{4}\right), \quad y > -\frac{1}{2}$$

$$25 + 2(x-3)^2 - 18 = 4\left(y + \frac{1}{2}\right)^2 - 1, \quad y > -\frac{1}{2}$$

$$2(x-3)^2 = 4\left(y + \frac{1}{2}\right)^2 - 8, \quad y > -\frac{1}{2}.$$

Then, moving $4(y + \frac{1}{2})^2$ to the left and dividing by -8 to get 1 on the right,

$$\frac{2(x-3)^2}{-8} - \frac{4\left(y + \frac{1}{2}\right)^2}{-8} = 1, \quad y > -\frac{1}{2}$$

$$\frac{\left(y + \frac{1}{2}\right)^2}{2} - \frac{(x-3)^2}{4} = 1, \quad y > -\frac{1}{2},$$

so we use $x = 3 + 2\sinh t$, $y = -\frac{1}{2} + \sqrt{2}\cosh t$ for $-\infty < t < \infty$.

21. Using the formula for $\sinh x$ with imaginary inputs, we have

$$\sinh(ix) = \frac{e^{ix} - e^{-ix}}{2}.$$

Substituting $e^{ix} = \cos x + i\sin x$ and $e^{-ix} = \cos x - i\sin x$, we have

$$\sinh(ix) = \frac{(\cos x + i\sin x) - (\cos x - i\sin x)}{2}$$

$$= i\sin x.$$

Solutions for Chapter 12 Review

Exercises

1. The coefficients of x^2 and y^2 are equal, so this is a circle with center $(0, 3)$ and radius $\sqrt{5}$.

5. Dividing by 36 to get 1 on the right gives

$$\frac{9(x-5)^2}{36} + \frac{4y^2}{36} = \frac{36}{36}$$
$$\frac{(x-5)^2}{4} + \frac{y^2}{9} = 1.$$

This is an ellipse centered at $(5, 0)$, with $a = 2, b = 3$.

9. One possible answer is $x = 3\cos t, y = -3\sin t, 0 \le t \le 2\pi$.

13. The ellipse $x^2/25 + y^2/49 = 1$ can be parameterized by $x = 5\cos t, y = 7\sin t, 0 \le t \le 2\pi$.

Problems

17. $x = t, y = 5$.

21. Rewriting as

$$2x^2 - 4x - y^2 + 2y,$$

factoring out 2 from $2x^2 - 4x = 2(x^2 - 2x)$ and -1 from $-y^2 + 2y = -(y^2 - 2y)$, and completing the square on $x^2 - 2x$ and $y^2 - 2y$ gives

$$2(x^2 - 2x) - (y^2 - 2y) = 0$$
$$2((x-1)^2 - 1) - ((y-1)^2 - 1) = 0$$
$$2(x-1)^2 - 2 - (y-1)^2 + 1 = 0$$
$$2(x-1)^2 - (y-1)^2 = 1.$$

This is a hyperbola centered at $(1, 1)$ with $2 = 1/a^2$, so $a = 1/\sqrt{2}, b = 1$, and opening left-right.

25. The plot looks like Figure 12.11.

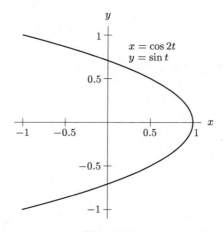

Figure 12.11

which does appear to be part of a parabola. To prove that it is, we note that we have

$$x = \cos 2t$$

$$y = \sin t$$

and must somehow find a relationship between x and y. Recall the trigonometric identity

$$\cos 2t = 1 - 2\sin^2 t.$$

Thus we have $x = 1 - 2y^2$, which is a parabola lying along the x-axis, for $-1 \le y \le 1$.

CHECK YOUR UNDERSTANDING

1. True. Since $x = \sin(\pi/2) = 1$ and $y = \cos(\pi/2) = 0$, the object is at $(1, 0)$.

5. False. There are many parameterizations; $x = \cos t, y = \sin t$ and $x = \sin t, y = \cos t$ are two of them.

9. True. Since

$$\frac{x^2}{4} + y^2 = \frac{(2\cos t)^2}{4} + (\sin t)^2 = \cos^2 t + \sin^2 t = 1,$$

these equations parameterize the ellipse $(x^2/4) + y^2 = 1$.

13. False. The correct identity is $\cosh^2 x - \sinh^2 x = 1$.

NOTES

NOTES

NOTES

NOTES

NOTES